# The Baofeng Radio Family Lifeline

The Real-World Protocol to Design a Mission-Ready Communication Plan, Master CHIRP Programming, Configure Repeater Offsets, and Train Your Household

MorseCode Publishing

# Contents

# THE PROMISE

**IN 14 DAYS, YOU will stop guessing and start communicating.**

If you follow the steps in this book, you will move from confusion to competence. You will:

1. **Build a working channel plan** that actually makes sense for your area, filtering out the digital noise and dead frequencies that clog up your memory banks.

2. **Program repeaters correctly** (so people can actually hear you). No more "chunking" the repeater without getting a response because you missed a tone setting.

3. **Run 3 real-world drills** to prove your gear works when cell service dies. You will know exactly how far you can talk in your specific neighborhood, through your specific trees and buildings.

4. **Troubleshoot failures calmly** instead of panicking when you hear static. When the radio acts up (and it will), you'll know if it's a battery issue, an antenna issue, or user error

You bought the radio because you wanted to be prepared. Right now, you just own a piece of plastic with too many buttons. Two weeks from now, you will own a **capability**. You will know exactly which button to press, even in the dark.

## WHO THIS BOOK IS FOR

- **The "Just in Case" Guy:** You bought a UV-5R off Amazon three years ago, threw it in a go-bag next to a first aid kit, and haven't touched it since. It's probably sitting next to a battery that hasn't been charged since 2021. You know you *should* learn how to use it, but the manual reads like stereo instructions.

- **The "Tactical Dad":** You aren't trying to play soldier. You just want a backup way to talk to your family during storms, power outages, or crowded events like the county fair when cell towers get jammed. You want a "Gray Man" setup, effective, discreet, and reliable.

- **The Frustrated Beginner:** You got your license (or are studying for it), but the classes only taught you electrical theory, not how to actually use the radio in your hand. You can hear the local repeater, but nobody ever answers when you call.

## WHO THIS BOOK IS NOT FOR

- **The Elitist:** If you want to argue about why cheap radios are ruining the RF spectrum or why everyone should buy a $400 Yaesu, put this back on the shelf. We know it's a cheap radio. That's the point. It's the radio people actually have.

- **The Cosplayer:** If you think buying the radio makes you an operator, you're in for a rude awakening. Hanging a radio on a plate carrier doesn't mean you can communicate. Skill matters more than gear.

- **The Pirate:** If you want to jam police frequencies, troll on GMRS channels, or act like an idiot on the air, we can't help you (and we won't). This book teaches responsible, effective, legal communication.

# HOW TO USE THIS BOOK (CHOOSE YOUR PATH)

You don't have to read this cover-to-cover right now. This is a field guide, not a novel. Jump to the problem that is making you want to throw the radio across the room.

- **"I can hear people, but nobody hears me."**

    - *Go to Chapter 3 (Repeater Mastery)* or *Chapter 1 (The 5 Settings That Kill You)*. We'll fix your offsets and tones in ten minutes.

- **"My computer won't talk to the radio (Driver/Cable Hell)."**

    - *Go to Chapter 4 (CHIRP 2.0)*. We'll bypass the bad drivers and get your spreadsheet loaded.

- **"I can't reach my house from the grocery store."**

    - *Go to Chapter 5 (Range Reality Lab)*. Physics is stubborn, but we can optimize your antenna and placement to squeeze every mile out of 5 watts.

- **"I want a plan for my family that isn't complicated."**

    - *Go to Chapter 9 (The Family Comms Plan)*. We'll turn "Here, take this" into a structured, stress-tested communication protocol.

**A Note on "The Gray Man" Approach** Throughout this book, we prioritize being **effective** over being loud. The goal isn't to look cool with a giant whip antenna and a tactical vest. The goal is to have a discrete, reliable lifeline when the phones go down. We follow the rules, we respect the airwaves, and we keep a low profile. The best operator is the one nobody notices until they need help.

# INTRODUCTION
## THE UPGRADE MINDSET

### The Driveway Moment

IT'S SATURDAY AFTERNOON. YOU'VE finally convinced your wife (or buddy) to help you test those radios you bought six months ago. The sun is shining, and you feel ready.

You hand her one unit and say, "Just stay in the kitchen. I'm going to drive to the end of the block. If things get weird, we'll use these."

You feel capable. Prepared. You drive three streets away, park, and key the mic.

*"Check, check. Can you hear me?"*

Silence.

You try again, speaking louder this time. *"Honey, radio check. Over."*

Static. Then, a weird garbled noise that sounds like a fax machine dying. Then, silence again.

You drive back to the house, frustrated, feeling the heat rise in your cheeks. You spent $60 on radios and $20 on a cable, and right now, shouting out the window would be more effective.

She hands you the radio and says, "I heard a noise, but I couldn't understand you. And then it started scanning by itself and picking up the weather station."

Congratulations. You have just experienced the **Baofeng Reality Check.**

Most people stop here. They assume the radio is junk (it's not), or that range is a lie (it's complicated), or that they just aren't smart enough to figure it out (false).

The problem isn't the radio. The problem is that you treated a professional tool like a toy walkie-talkie. You expected it to work like an iPhone, you push a button and it just works. But RF (Radio Frequency) doesn't care about your expectations.

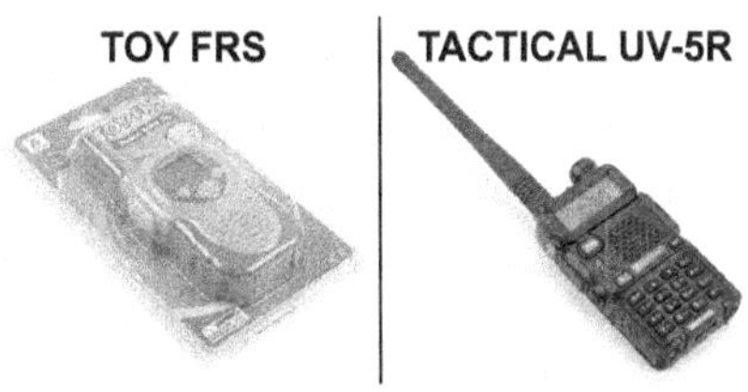


## The 4 Truths of The Baofeng Radio Family Lifeline

To move from "guy with a radio" to "capable operator," you have to accept four truths. These are the pillars of the 2.0 mindset:

**1. Expectations: It is a $25 Radio.** It has a slow scanner. The receiver filters noise about as well as a screen door holds back water. It is not a Motorola public safety radio. But if you understand its quirks and work around them, it is 90% effective for 10% of the price. That is a trade we will take every time.

**2. Practice is a Perishable Skill.** You cannot learn CHIRP software during a blackout. You cannot figure out "Simplex vs. Duplex" while a hurricane is rattling your windows. Adrenaline dumps your IQ by about 20 points. If you have to look up the manual to change a frequency, you are dead in the water. You verify *now* so muscle memory takes over *later*.

**3. Planning > Frequencies.** Having 100 random channels programmed in your radio is useless. A list of repeater frequencies for a state you'll never visit isn't preparedness; it's digital hoarding. Having 5 channels that you and your family know by heart, and act on instinctively, is a superpower.

**4. Diagnostics is the Skill.** Things will break. Batteries die. Settings get bumped in your bag. Connectors corrode. The "2.0 Operator" doesn't panic; they troubleshoot. This book teaches you how to look at a silent radio, run a 60-second diagnostic, and fix it in the field.

### The Operator Upgrade Ladder

This book is designed to move you up this ladder. Where are you right now?

- **Level 1: The Listener (Passive).** You can turn it on and hear the weather channel. You are essentially holding a police scanner that is hard to program. You have zero transmit capability.

- **Level 2: The Programmer (Technical).** You can use software to put channels in. You have a spreadsheet, but you don't really know if those channels work. You are technically set up, but operationally untested.

- **Level 3: The Communicator (Active).** You have made contact. You know your range limits. You can hit a repeater and get a signal report. You trust the gear because you've used it.

- **Level 4: The Troubleshooter (Resilient).** When it fails, you know why. You have a backup plan (and a backup battery). You are the "comms guy" for your family or group.

## A Note on Hardware (The "Hero" Device)

In this book, we use the classic **Baofeng UV-5R** for our diagrams and instructions. It's the Honda Civic of radios, ubiquitous, cheap, customizable, and capable.

**VARIANT CALLOUT: "Is my radio different?"** If you own a **UV-82** (the one with the rocker switch), a **GMRS-V1**, or a **UV-5G**, your radio is 95% the same as the UV-5R inside. However, you might have locked menus (due to FCC rules) or different buttons. Look for these **Variant Callout** boxes throughout the book. We will

tell you exactly when your specific model requires a different step so you never feel left behind.

## ACTION STEP: The 10-Minute Assessment

Before you turn the page, put your radio on the table. Be honest with yourself. We need to establish a baseline before we can upgrade.

**Your Mission Checklist:**

- **Power:** Does the battery hold a charge? (If it's been in a drawer for a year, charge it *now*. A dead radio is just a paperweight).

- **Connection:** Do you have the programming cable? (Not just the charging cable, the USB cable that connects to your PC. You cannot act as a Level 2 Operator without this).

- **Legal:** Do you have your GMRS license or Ham Technician license? (If not, we aren't the police, but we strongly suggest getting legal. It protects the hobby and lets you practice without looking over your shoulder. GMRS is the "easy button", $35, no test, covers the whole family. Go do it.)

- **The "Baseline":** Turn the radio on. Press the VFO/MR button (orange button on many models). Does the radio speak "Channel Mode" or "Frequency Mode"?

If you don't know what that last bullet point means, don't worry. It's the single most confusing button on the device, and we're going to master it in Chapter 1.

Let's calibrate.

Jared Johnson (KF0RTU)

Jared@MorseCodePublishing.com

Chapter 1

# THE 2.0 CALIBRATION

## WHAT MATTERS NOW

You have the radio in your hand. Maybe you've already mashed some buttons and heard static. Maybe it's fresh out of the box and still smells like factory plastic.

Before we program a single frequency, we need to calibrate the device, and more importantly, calibrate *you*.

Most people skip this step. They tear open the box, Google "police frequencies," and try to program their local dispatch immediately. Then they get frustrated when the radio beeps at them, refuses to transmit, or transmits on a frequency that interferes with actual emergency services.

We are going to strip this down to the "Known Good" state. We are building a foundation so solid that when things go sideways in the dark, when the power is out, the rain is hammering the roof, and your family is looking to you for answers, you know for a fact that it's not the radio's settings that are fighting you.

## THE 5 SETTINGS THAT CAUSE 80% OF THE PAIN

If you understand these five concepts, you understand the Baofeng. If you ignore them, you will forever be "the guy with the radio that doesn't work," frantically pressing buttons while everyone else is communicating.

### 1. VFO vs. MR (The Two Brains)

Your radio has a split personality. It has two distinct operating modes, and you must know which brain is in charge at all times. If you don't, you will try to save a channel and fail, or try to tune a frequency and get locked out.

**Frequency Mode (VFO - Variable Frequency Oscillator):** This is the **manual** brain. It is your scratchpad. You type in numbers directly (e.g., 1-4-6-5-2-0). The radio goes there immediately.

*When to use it:* When you are searching for a new signal, testing a frequency you just found online, or meeting a buddy on a random frequency that isn't in your saved list.

- *Behavior:* Changes here are temporary unless you explicitly save them to memory.

- **Channel Mode (MR - Memory Recall):** This is the **saved** brain. It is your contact list. You cannot change the frequency numbers here; you can only scroll through what you have previously saved (Channel 1, Channel 2, etc.).

  - *When to use it:* In 99% of real-world scenarios. Under stress, you don't want to type 4-6-2...; you just want to click to "Channel 1: HOME."

  - *Behavior:* This mode protects you from accidentally bumping the keypad and changing your frequency.

**The Trap:** You try to type a specific frequency while in Channel Mode. The radio ignores you, or worse, it interprets your button presses as a command to jump to "Channel 146," which is empty. You think the keypad is broken. It's not; you're just in the wrong brain.

**The Fix:** The VFO/MR button (orange button on the UV-5R) toggles between them.

- **Visual Check:** Look at the right side of the display.

  - If you see small channel numbers (001, 002, 127), you are in **Channel Mode.**

  - If there are *no* small numbers on the right, you are in **VFO Mode.**

- **Audible Check:** The radio literally announces "Frequency Mode" or "Channel Mode" when you press it (unless you turned the voice off, which we'll discuss later).

## 2. Duplex/Offset (The Repeater Math)

Simplex means you talk and listen on the exact same frequency (like a normal walkie-talkie or FRS radio). Repeaters are different. They are "deaf" on the frequency they broadcast on. To talk to a repeater, you must listen on one frequency but transmit on another (usually 0.6 MHz away for Ham, or 5.0 MHz for GMRS).

**The Logic:** Imagine a translator standing on a hill. He shouts loudly to everyone in the valley (Output Frequency). But to get him to shout your message, you have to whisper in his ear (Input Frequency). If you shout at his mouth, he won't hear you. The "Offset" is simply the distance between his mouth and his ear.

**The Trap:** You program a repeater frequency, but you forget the offset. You press the PTT (Push-to-Talk), and you transmit on the listening frequency (shouting at the translator's mouth).

- **Symptom:** You can hear people talking clearly. When you transmit, nobody responds. You might hear a "kerchunk" (the repeater activating) but your audio doesn't go through.

**The Fix:**

- **Menu 25 (SFT-D):** Direction of the shift.

  - **OFF:** Simplex (Direct radio-to-radio).

  - **+ (Plus):** Transmit frequency is *higher* than receive.

  - **- (Minus):** Transmit frequency is *lower* than receive.

- **Menu 26 (OFFSET):** The amount of shift.

  - Standard Ham (2m): 00.600 (Type all five digits)

- ○ Standard GMRS: 05.000

- ○ Standard Ham (70cm): 05.000

## 3. Tone Logic (The Key to the Door)

Think of a standard frequency as a house. Anyone can stand on the sidewalk and yell (transmit). A "Tone" (CTCSS or DCS) is a lock on the door.

- **Encode Tone (T-CTCS / Menu 13):** This puts a "key" in your pocket. When you speak, your radio sends a sub-audible tone (invisible key) along with your voice. The repeater (the house) checks for this key. If you have it, it opens the door and broadcasts your message. If you don't, it ignores you. **Most repeaters require this.**

- **Decode Tone (R-CTCS / Menu 11):** This puts a lock on *your* speaker. Your radio will stay 100% silent, even if the frequency is busy, unless the incoming signal has the matching key.

**The Trap:** You set a "Receive Tone" (R-CTCS) when you didn't mean to. This is the "Cone of Silence" error.

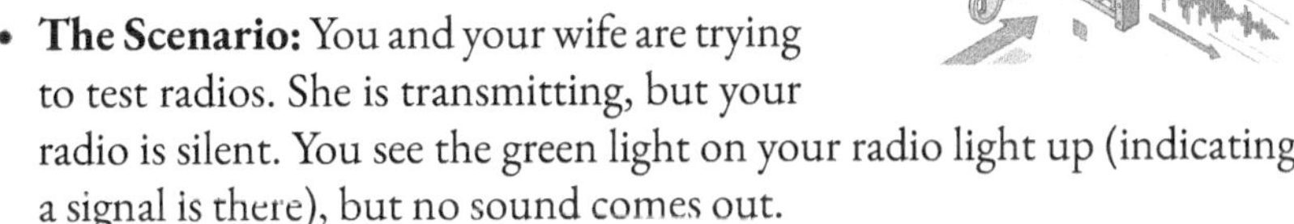

- **The Scenario:** You and your wife are trying to test radios. She is transmitting, but your radio is silent. You see the green light on your radio light up (indicating a signal is there), but no sound comes out.

- **The Cause:** Your radio is waiting for a specific tone key that she isn't sending. You locked your own door and didn't give her the key.

**The Golden Rule:** Unless you have a specific reason (like living near a interference-heavy area or blocking out other repeater traffic), **leave R-CTCS (Menu 11) and R-DCS (Menu 10) set to OFF.** Only set the **T-CTCS (Menu 13)** when a repeater requires it.

## 4. Scan Behavior (CO / TO / SE)

You will want to scan channels to monitor traffic, maybe to listen for storm spotters or find activity during an emergency. But *how* the radio stops scanning determines if you catch the info or miss it entirely.

**The Trap:** You set it to scan. The radio lands on a conversation, maybe a storm report. "Tornado spotted near..." and then suddenly, the radio starts scanning again, cutting off the location. Why? Because you told it to move on after 5 seconds, regardless of whether they were done talking.

**The Fix (Menu 18 - SC-REV):**

- **TO (Time Operation):** The "Sample" Mode. It stops for 5 seconds to let you hear a snippet, then moves on no matter what. **Terrible for following conversations.**

- **CO (Carrier Operation):** The "Listening" Mode. It stops while someone is talking. It stays there as long as the signal is present. It resumes scanning only after the signal drops for a few seconds. **This is what you want for general monitoring.**

- **SE (Search Operation):** The "Find It" Mode. It stops when it finds a signal and *stays there*. It stops scanning completely. Use this when you are hunting for active channels and want to lock onto the first one you find.

- **Action:** Go to Menu 18 and set this to **CO** right now.

## 5. Squelch Realities (The Gatekeeper)

Squelch is the noise gate. It decides how strong a signal must be before the speaker turns on.

- **Too Low (0):** The gate is open. You hear constant static (white noise).

- **Too High (9):** Only a massive signal (someone standing next to you) will open it. You will miss weak signals from distant repeaters.

**The Trap:** The factory settings on Baofengs are notoriously bad. The difference between Level 1 and Level 9 is often negligible on stock firmware. They don't scale linearly.

**The Fix (Menu 0 - SQL):**

- Set this to **1** for now. (Note: In a dense city with lots of interference, you might need **3** or **5** to stop the static pops.) It's the most sensitive setting that isn't constant static.

- *Pro Tip:* If you are driving and hear constant "PFFT... PFFT..." static pops, bump it to 2 or 3. If you leave it at 5 or 6, you might miss a weak emergency transmission.

# YOUR BASELINE: CREATE A KNOWN-GOOD STARTING POINT

We are going to wipe the slate clean. Out of the box, these radios often come with random Chinese test frequencies, marine channels, or interference-prone garbage programmed into the memory. We don't want those ghost channels.

## The Reset Strategy

*Warning: This deletes all saved channels. If you bought a customized "Pre-Programmed" radio from a vendor like BetterSafeRadio or Rugged Radios, **SKIP THIS STEP**. You paid extra for that programming; don't wipe it out. Just do the "Settings Check" below.*

**VARIANT CALLOUT: GMRS & Locked Radios** If you have a **UV-5G, GMRS-V1, or UV-82C**, your radio firmware may block the "Reset All" function to comply with FCC Part 95 rules (you cannot delete the primary GMRS channels). That's fine. Just manually verify the settings in the Menu using the worksheet below.

## The "Tactical Dad" Factory Reset (Standard UV-5R):

1. Turn radio ON.

2. Press Menu, type 40 (RESET).

3. Press Menu, select ALL.

4. Press Menu again to confirm. The radio will go black, reboot, and likely speak a Chinese greeting.

5. **Don't panic.** We anticipated this.

6. Press Menu, type 14 (VOICE), press Menu. Use the arrow keys to select ENG (English). Press Menu to save. Now it speaks your language.

### The "Baseline Channel" (Channel 000)

We are going to program one channel manually right now using the keypad. **Pro

**Tip:** Type fast. If you pause for more than 5 seconds, the radio gets bored and exits the menu. This is your "sanity check" channel. If you ever get confused, lost in menus, or think the radio is broken, you go to Channel 000.

**Your Mission:** Find the National Simplex Frequency for your service. This is the "Town Square" of radio frequencies.

- **Ham:** 146.520 MHz

- **GMRS:** 462.5625 MHz (This is universally Channel 1 on GMRS radios)

**Why?** This is the common meeting place. If you are lost, you go here. If you are testing, you go here. If you are trying to make contact with a stranger in an emergency, this is your best bet.

## REALITY CHECK: WHAT THE BAOFENG IS (AND ISN'T)

Before we start building your plan, let's be honest about the hardware. Managing expectations is the key to happiness.

**What it is GREAT at:**

- **Disposable Utility:** If you drop it in a creek, step on it, or hand it to a neighbor who loses it, you are out $25. This makes you willing to actually *use* it. You won't hesitate to toss it in a glovebox or a backpack.

- **Battery Life:** With a decent battery (and not scanning constantly), these will sit on standby for days. The idle power consumption is very low.

- **Versatility:** It listens to FM radio, Marine bands, NOAA weather, and more. It is a Swiss Army Knife.

**What it SUCKS at:**

- **The "Front End" (Receiver Quality):** This is the biggest weakness. The Baofeng receiver chip is easily overwhelmed by strong signals.

  - *The Scenario:* You are driving near a cell tower or a high-power paging transmitter. Suddenly, your radio goes silent. Or maybe it blasts static.

  - *Desense:* The radio is "desensitized." It's like trying to hear a whisper while someone is screaming in your other ear. The radio protects itself by going deaf.

- ○ *The tell:* If you are in the city and can't hear anything, unscrew the antenna. If you suddenly hear a split second of clear audio before the signal drops, your radio was overloaded.

- **Scan Speed:** It is slow. Painfully slow. Do not try to scan 100 channels. By the time it cycles through, the conversation will be over. Keep your scan lists short (15-20 channels max).

- **Stock Antenna:** It's often called a "rubber ducky." It is durable, but electrically, it is a compromise. It works for 1-2 miles, but don't expect miracles. (We fix this in Chapter 5).

## ACTION STEPS

### 1. The "Before You Troubleshoot" Checklist

*Copy this onto an index card, laminate it with packing tape, and tape it to the back of your battery.*

If the radio isn't working, check these in order:

1. **Battery:** Is the voltage over 7.4V? (Hold the 0 key for 2 seconds to check voltage. On some older models, hold the SCAN key instead). A low battery causes weird glitches before it actually dies.

2. **Mode:** Are you in VFO or MR? (Press VFO/MR). If you are trying to change channels, be in MR. If you are typing frequencies, be in VFO.

3. **Tone:** Do you see "CT" or "DCS" on the left of the screen? If you are trying to talk to a new radio and you see these letters, turn them OFF. You are likely locking yourself out.

4. **Offset:** Do you see a little + or - at the top of the screen? If you are trying to talk Simplex (radio-to-radio), this MUST be off.

5. **Antenna:** Is it actually screwed on tight? (Connector SMA-Female vs Male confusion is real, make sure the center pin actually connects).

## 2. My Baseline Settings Worksheet

*Go into your menu and set these NOW. This is your foundation. Do not pass Go until these match.*

| Menu # | Name | Setting | Why? |
|---|---|---|---|
| 0 | SQL | 1 | **Max sensitivity.** We want to hear everything. If it's too noisy, bump to 2. |
| 1 | STEP | 2.5K or 5.0K | **Precision.** Allows you to tune to specific narrowband frequencies without rounding errors. |
| 2 | TXP | HIGH | **Power.** Default to max power. You can always switch to Low (tap the # key) to save battery later. |
| 3 | SAVE | OFF | **CRITICAL.** Battery save mode sleeps the receiver. It causes you to miss the first 1-2 seconds of a transmission (the "Baofeng delay"). Turn it OFF for reliability. |
| 7 | TDR | OFF | **Focus.** Dual Watch lets you listen to two channels. It is confusing for beginners. Keep it OFF so you know exactly who you are talking to. |
| 14 | VOICE | ENG | **Feedback.** Helps you know what mode you are in without looking. |
| 18 | SC-REV | CO | **Carrier Operation.** Ensures the scanner resumes only after the signal drops, so you don't miss the end of the sentence. |
| 25 | SFT-D | OFF | **Simplex Default.** We start with no offset. We add offsets only when programming specific repeaters. |
| 39 | ROGER | OFF | **Etiquette.** The "Roger Beep" screams, "I am a newbie with a cheap radio." It is annoying to other operators. Turn it off. |

**Once these are set, your radio is calibrated. You are now a Level 1 Operator. You have a known-good device. Turn the page.**

# Chapter 2

# BUILD YOUR CHANNEL PLAN

## BEFORE YOU TOUCH CHIRP

MOST PEOPLE PROGRAM THEIR radios like they pack for a vacation: they throw everything they might possibly need into the suitcase until the zipper bursts.

They go to a database website, download a CSV file of "All Repeaters in the State," upload 127 random frequencies into the radio, and call it good. They feel prepared because their memory banks are full.

**This is a tactical disaster.**

Data hoarding is not preparedness. When you are stressed, cold, or driving in the rain with a power outage looming, you cannot scroll through 80 channels labeled K5PLA, W3ABC, and N5XYZ trying to remember which one covers your neighborhood and which one is three counties away.

You will scroll past the one you need, or you will be scanning so much junk that you miss the urgent call from your wife. You will be stuck listening to a repeater 40 miles away while your local team is trying to reach you on Simplex.

Before we open the software or connect a cable, we are going to grab a piece of paper and design a **System**. We are going to build a mental map that works even when you are tired and confused.

# THINK LIKE A SYSTEM DESIGNER

The Baofeng UV-5R is dumb hardware. It does not have "Memory Banks," "Groups," or "Zones" like a $5,000 Motorola public safety radio. It just has a big, disorganized bucket of 128 channels (000–127).

If you mix Simplex channels, Repeaters, Marine bands, and NOAA weather stations randomly in that bucket, you have created a mess. We need to impose order on chaos. We need to turn that bucket into a filing cabinet.

## 1. The "Virtual Zone" Strategy

Professional radios have a physical knob to switch between "Zones" (e.g., Police Zone, Fire Zone, Tactical Zone). Since the Baofeng doesn't have that knob, we have to create **Virtual Zones** using channel number ranges.

By reserving specific blocks of numbers for specific tasks, you reduce the mental load required to find a frequency.

- **Zone 1 (Channels 000–019): The "Hot" Zone.**

    - These are the channels you use 90% of the time. This includes your Simplex frequencies, family channels, and your one primary repeater.

    - *The Tactical Advantage:* You can access these by typing 0-0-1 or 0-1-0 instantly. Muscle memory works here. If you are in the dark, you know the low numbers are the safe numbers.

- **Zone 2 (Channels 020–059): The Repeater Bank.**

    - All the secondary repeaters in your county and the surrounding counties.

    - Organize these geographically (e.g., North repeaters first, then South) or by priority. These are the channels you switch to when the "Hot Zone" repeater is down or out of range.

- **Zone 3 (Channels 060–089): Interoperability (Interop).**

    - MURS, GMRS, FRS (if you are scanning them). The channels "other people" use.

- ○ If you hear a neighbor calling for help on a Walmart blister-pack radio, they will likely be here. You keep these separate so they don't clutter your main operation.

- **Zone 4 (Channels 090–127): The Boneyard.**

  - ○ Weather stations, Police/Fire listen-only frequencies, and travel channels.

  - ○ These are "Read Only" channels. You put them at the end so you don't accidentally scroll past them when looking for your family channel.

**Why?** Because if you need to talk to your family, you know instinctively to type 0-0-1. You don't have to hunt. You don't have to look at the screen. You just know.

## 2. Naming Conventions Under Stress

The UV-5R display gives you **7 characters**. That is it. If you name a channel NorthMountain, the radio will likely truncate it to NorthMo or just show the frequency numbers.

We need a naming language that tells you **Who** and **What** at a glance. Under stress, your IQ drops. You won't remember that 146.880 is the "Club Repeater." You need the screen to tell you.

**The "Bad" Way:**

- K5ABC (This is a callsign. Who is it? Where are they? Is this a repeater or a person?)

- RPTR 1 (Which repeater? The one near your house or the one near work?)

- 146.520 (Just the raw frequency. Unless you are a math savant, this means nothing in an emergency.)

**The "Tactical Dad" Way:**

- HOME-S (Home Simplex. It tells you exactly who it is for and that it is Simplex.)

- CONVOY (Car-to-Car. Universal channel for travel.)

- GMRS-15 (Standard GMRS Ch 15. Standardized naming helps when talking to others.)

- CITY-R (City Repeater. Tells you the location and the type.)

- SHRF-L (Sheriff Listen-Only. Reminds you *not* to transmit.)

**The Spouse Test:** If you hand the radio to your spouse or teenager and say, "Go to the Home channel," can they figure it out? If the screen says 146.520, probably not. If it says HOME-S, yes. The radio should speak human, not robot.

## THE 3-LAYER PLAN

We are going to fill your "Virtual Zones" using a three-layer approach. This ensures you have layers of redundancy. If Layer 2 fails, you fall back to Layer 1. If Layer 1 is out of range, you try Layer 3.

### Layer 1: Simplex Local (The Lifeboat)

**Reliability: High | Range: Low (1–3 miles)**

This is direct radio-to-radio communication. It relies on zero infrastructure. It works when repeaters are down. It works when power is out. It works when you are hiking in a dead zone. It is the only layer you can 100% control.

- **Channel 001:** CALL-S (National Calling Freq: 146.520 Ham or 462 .5625 GMRS). This is the "Town Square."

- **Channel 002:** HOME-S (Your chosen family frequency). Pick a quiet frequency away from the National Calling channel. **Pro Tip:** This is the ONE place where adding a Tone (CTCSS) is smart. If you set both radios to use Tone 100.0Hz, you won't hear the random trucker driving by on the same frequency. It creates a "semi-private" line for your house.

- **Channel 003:** TAC-1 (A secondary frequency).

  - *Scenario:* You are coordinating a convoy, but HOME-S is crowded with chatter from another group. You simply say, "Switch to Tac 1," and everyone clicks one click up.

**Variant Callout:** *If you are using GMRS, you are legally restricted to specific channels. Use Channel 15–22 for this layer. If you are Ham, look up the "Band Plan" for your area to find a dedicated simplex frequency (e.g., 146.580).*

## Layer 2: Primary Repeaters (The Grid)

### Reliability: Medium | Range: High (10–50 miles)

These are the big towers on mountains. You use them to talk across the county. However, repeaters are owned by people. They can lose power. They can break. They can be shut down. Never rely on them as your *only* option.

- **Selection Criteria:** Do not program every repeater in the state. Pick the **Top 3**:

    a. The one that covers your **House**.

    b. The one that covers your **Work**.

    c. The one used by the local **ARES/RACES** (emergency response) group.

- **Naming:** CNTY-R (County Repeater), ARES-P (Emergency Net Primary).

## Layer 3: Travel/Backup (The "Just in Case")

### Reliability: Low | Range: Variable

This is the stuff you scan just to know what's going on. Information gathering is just as important as transmitting.

- **NOAA Weather:** Program your local NOAA frequency. (You will set this to *Skip Scan* later, but you need it in the memory).

- **Local EMS/Fire:** (If they aren't encrypted/digital). Hearing the fire dispatch can tell you where the storm damage is before the news does. Check RadioReference.com for your county. If the Mode column says "FM" or "NFM", you can listen. If it says "P25", "DMR", or "NXDN", your Baofeng is unfortunately deaf to it.

- **MURS/FRS:** The "General Public" frequencies. Good for listening to see if anyone nearby is calling for help.

# SCAN STRATEGY THAT DOESN'T BETRAY YOU

Scanning on a Baofeng is slow. It scans about 3 channels per second. Compare that to a police scanner that does 50+ channels per second.

If you scan all 128 channels, it takes **40+ seconds** to loop around once. A typical radio transmission ("This is Base, checking in.") takes 3 seconds. You will mathematically miss 90% of the conversation.

**The "Exclude" Tactic**

In Chapter 4, we will use CHIRP to set "Skip" flags. But for your plan right now, mark which channels you actually want to scan.

**The Golden Rule of Scanning:** Only scan channels you are willing to answer.

- **Do Scan:** Your Home channel, the main Repeater, the National Calling frequency. Keep this list tight, under 15 channels if possible.

- **Do Not Scan:** The 50 other repeaters you rarely use.

- **The "NOAA Trap":** Never, ever add the NOAA weather channel to your scan list. It transmits a continuous signal 24/7. Your scanner will hit it, stop, and stay there forever. You will never hear anything else.

**Priority Scan vs. Dual Watch (TDR)**

You have two ways to "watch" your home channel while listening to others.

1. **Dual Watch (TDR):** The radio listens to Channel A (Top Line) and Channel B (Bottom Line) at the same time, switching back and forth rapidly.

   - *Pro:* Great if you only care about two channels (e.g., Home + Repeater). It is faster than scanning.

   - *Con:* It can be confusing to know which one you are transmitting on if you aren't paying attention to the little arrow on the screen.

2. **Scanning:** The radio cycles through a list of enabled channels.

   - *Pro:* Covers more ground (Home, Repeater, Interop, and Fire Dispatch).

○ *Con:* You might miss the start of a transmission because the radio was busy checking a quiet channel.

**Recommendation:** For daily carry, use **Dual Watch**. Set Top Line (A) to your primary Repeater (for situational awareness). Set Bottom Line (B) to your Home Simplex (for family). This guarantees you never miss a call from home.

# WORKBOOK TOOLS

### Action Step: Fill Out Your Paper Plan

*Do not skip this. Write it down before you touch the computer. Having this on paper is your backup when the computer crashes.*

### The "Tactical Dad" Channel Worksheet

| Ch # | Name (7 Char Max) | Freq | Tone? | Purpose | Scan? |
| --- | --- | --- | --- | --- | --- |
| 000 | CALL-S | 146.520 | None | National Call (Ham) | Yes |
| 001 | HOME-S | | | Family Primary | Yes |
| 002 | TAC-1 | | | Backup / Hiking | No |
| 003 | MURS-3 | 151.940 | None | Public Interop | Yes |
| 010 | **CNTY-R** | | 103.5 | Main County Rptr | Yes |
| 011 | SKYWRN | | | Weather Spotters | Yes |
| 020 | NOAA | 162.550 | None | Weather Audio | **NO** |
| 127 | POLICE | 155.xxx | None | Listen Only | No |

### Your Assignment:

1. Identify your **1 Simplex Frequency** for the family.

2. Identify your **2 Primary Repeaters** (Check RepeaterBook.com).

3. Write them in the chart.

4. Create 7-character names for them.

Once this is written down, programming in Chapter 4 will take 5 minutes. If you try to decide while programming, it will take 5 hours.

# Chapter 3

# REPEATER MASTERY

## PROGRAM IT. VERIFY IT. TRUST IT.

YOU'VE STOOD IN YOUR driveway, keyed the mic on 146.520, and heard nothing but static. You walked to the end of the block, tried again, and got the same result. You realized the hard way that 5 watts of power doesn't go very far through suburban sprawl, hills, and pine trees.

Enter the **Repeater**.

A repeater is simply a massive "force multiplier" for your radio. It is a high-power radio installed on top of a mountain, a skyscraper, or a water tower. It listens for your weak, scratchy signal on one frequency and simultaneously screams it out on another frequency to the entire county.

It turns your 2-mile "line of sight" range into a 50-mile "county-wide" footprint. It is the difference between talking to your neighbor and coordinating with a family member three towns over.

But repeaters are picky. They are not open channels like FRS walkie-talkies. They are like a secure facility: if you don't knock on the door correctly (Tone) and enter through the right gate (Offset), they will ignore you completely. You can scream at them all day, and they won't even wake up.

This chapter turns "I think I hit the repeater" into "I have solid copy." We are going to move from guessing to knowing.

## FIND REPEATERS THE RIGHT WAY

Don't guess. Don't scan randomly hoping to find one; you'll mostly hear digital noise or interference. You need actionable intelligence.

**The Source:** Download the **RepeaterBook** app (free) or go to RepeaterBook.com. It is the gold standard database for amateur radio repeaters.

**The Data You Need (Write this down):** To successfully hit a repeater, you need four specific pieces of intel. If you are missing one, it won't work. It's like a combination lock, miss one number, and the shackle stays closed.

1. **Output Frequency (The "Listen" Freq):** This is the big number listed in the app (e.g., 146.880). This is what you listen to. It's the voice of the repeater.

2. **Offset (The Math):** The repeater cannot listen and talk on the exact same frequency at the exact same time (it would deafen itself). It listens on one frequency and talks on another. The "Offset" is the distance between them.

   ○ 2 Meter Ham is usually 0.600 MHz.

   ○ 70cm Ham is usually 5.000 MHz.

   ○ GMRS is always 5.000 MHz.

3. **Direction (The Shift):** Do you add (+) or subtract (-) the offset to find the input frequency? The app will tell you.

   ○ *Example:* If Output is 146.880 and Offset is -0.600, your radio will transmit on 146.280.

4. **Tone (The Key):** Look for "Tone," "PL," or "CTCSS." If it says 103.5, that is your electronic key card. You must program your radio to send this sub-audible tone. If you don't send it, the repeater ignores you. If it says "Open" or "CSQ" (Carrier Squelch), there is no lock, and you don't need a tone.

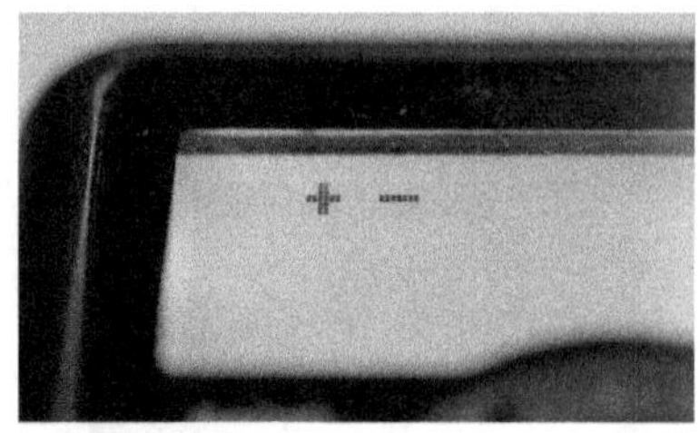

# PROGRAM REPEATERS TWO WAYS

We will start with the "Easy Way" (Software) to get your bulk list done, but you must master the "Rescue Method" (Keypad) for the field. If you are in a blackout, CHIRP won't save you.

## Method 1: The CHIRP Method (Desk Duty)

*We cover the full CHIRP workflow in Chapter 4, but here are the columns that specifically matter for repeaters. Getting these wrong is the source of 90% of forum posts asking for help.*

1. **Frequency:** Enter the **Output** frequency (e.g., 146.880). Do not enter the input frequency here.

2. **Duplex:** Set to + or - (matches the Direction). If you leave this as "(None)", you are programming a Simplex channel and the repeater won't hear you.

3. **Offset:** Enter the amount (e.g., 0.600). Double-check your decimal points. 6.000 will put you totally out of band.

4. **Tone Mode:** This is the tricky one.

   - If the repeater needs a tone: Set to Tone. This turns on the **Encode** (Transmit) tone.

   - If no tone: Set to (None).

   - *Warning:* Do NOT select ToneSql unless you are 100% sure you need it. ToneSql sets both Transmit AND Receive tones. If the repeater doesn't transmit a tone back (many don't), your radio will stay silent, and you won't hear the reply.

5. **Tone:** Enter the value (e.g., 103.5).

  ○ *Note:* The "ToneSql" column is ignored by the radio if "Tone Mode" is set to just Tone.

## Method 2: The Manual "Rescue Method" (Field Duty)

It's 2 AM, the power is out, trees are down, and you need to hit a repeater you didn't plan for. You don't have your laptop. You have your thumbs and a dim backlight.

**The "Tactical Dad" Keypad Sequence (UV-5R):** *Follow this exactly. If you pause for 5 seconds, the menu times out and drops you back to the main screen. Speed counts.*

1. **Go to VFO Mode:** Press VFO/MR until the radio speaks "Frequency Mode" and you see frequency numbers (no channel numbers on the right).

2. **Select A-Band:** Press EXIT/AB so the little black arrow is on the top line. Always program on the top line to avoid confusion.

3. **Enter Frequency:** Type the Output Freq directly (e.g., 1-4-6-8-8-0).

4. **Set the Tone (The Key):**

  ○ Press MENU, type 13 (T-CTCS), press MENU.

  ○ Use the Up/Down arrows to find your tone (e.g., 103.5).

  ○ Press MENU to save (Verify the arrow drops to the next line). Press EXIT.

  ○ *Why T-CTCS?* Because we only want to *transmit* the tone. We want to listen to everything.

5. **Set the Offset Amount (The Distance):**

  ○ Press MENU, type 26 (OFFSET), press MENU.

  ○ Type 00600 (for 2m) or 05000 (for 70cm/GMRS).

  ○ Press MENU to save. Press EXIT.

6. **Set the Offset Direction (The Shift):**

   ○ Press MENU, type 25 (SFT-D), press MENU.

   ○ Select + or - using the arrows.

   ○ Press MENU to save. Press EXIT.

7. **Save the Channel:**

   ○ Press MENU, type 27 (MEM-CH), press MENU.

   ○ Use arrows to find an empty channel (e.g., 020). Note: If the channel number has CH- in front of it (like CH-020), it is full. **The radio will not overwrite it.** You must first go to **Menu 28 (DEL-CH)** to delete it, then come back here to save.

   ○ **Crucial Step:** Press MENU to confirm. The voice says *"Receiving Memory."*

   ○ Press EXIT.

8. **Switch to Channel Mode:** Press VFO/MR. Use arrows to go to Channel 20.

**The "Quick Check":** While on the channel, tap the SCAN (*) button quickly (don't hold it). You should see an R appear on the screen, or the frequency number should shift briefly (e.g., from 146.880 to 146.280). If the numbers shift, your radio is doing the math correctly.

## VERIFICATION PROTOCOL

Just because it's programmed doesn't mean it works. I have seen men carry "programmed" radios for years that were incapable of transmitting because of a simple typo in the offset.

**1. Listen First (The 15-Minute Rule)**

Before you talk, listen. Amateur radio isn't a telephone; it's a party line.

• Is there a "Net" (scheduled meeting) happening?

• Are two people deep in a conversation about their hip surgery?

- Don't barge in. Wait for a break. If you interrupt a directed emergency net, you become the problem.

## 2. The "Kerchunk" Test

This is the non-verbal handshake.

- **Action:** Tap your PTT button for **one second** and release.

- **Listen:** Do you hear a burst of static, a "click," or a "beep-boop" courtesy tone return immediately after you let go?

- **Result:** That return sound is the "Squlech Tail." It means the repeater heard your signal, verified your tone, opened its receiver, and stayed on for a second before shutting down.

  - *If you hear the tail:* You are making it into the machine. Your Tone and Offset are correct.

  - *If silence:* You did not hit it. Check your settings or move to higher ground.

- **Etiquette:** Don't do this constantly. Better yet, key up and say **'Testing'** instead of just clicking. It keeps the cranky old-timers happy and is technically required by the rules.

## 3. The Signal Report (Radio Check)

Once you verify the repeater wakes up, ask for a human check. A repeater might hear your carrier wave (the kerchunk) but your audio could be too quiet or full of static.

**The Script:** "This is 'Your Callsign/Name', testing a new radio setup. Can anyone give me a signal report?"

**The Response (Decoding what they say):**

- **"Full Quieting":** This is the gold standard. It means your signal is so strong it is fully "quieting" the background static. You sound like you are standing next to them.

- **"Scratchy but readable":** You are understandable, but there is background hiss (bacon frying sound). You are likely on the edge of the repeater's range. Don't move while talking.

- **"You're into the machine, but no audio"**: You hit the repeater (the light went green), but your mic gain is too low, you are whispering, or your headset is broken. Speak closer to the mic.

- **"Picket Fencing"**: You are moving (driving or walking), and your signal is chopping in and out rapidly. Stop moving.

## COMMON FAILURE PATTERNS

### 1. The "Silent Treatment" (Wrong Tone Mode)

- *Symptom:* You transmit, the repeater wakes up (you hear the tail), but nobody answers you even though you hear them talking.

- *Cause:* You likely set **R-CTCS** (Receive Tone) on your radio. This puts a lock on *your* speaker. If the repeater output doesn't have a matching tone (and many don't), your radio keeps the speaker muted. You are deaf to the world.

- *Fix:* Turn **R-CTCS** (Menu 11) to OFF. Only use **T-CTCS** (Menu 13).

### 2. The "Wrong Door" (Wrong Offset Direction)

- *Symptom:* You transmit, and absolutely nothing happens. No "ker-chunk," no tail, no angry feedback.

- *Cause:* You set + instead of -. You are transmitting on 147.480 instead of 146.280. You are knocking on the back wall of the building instead of the front door.

- *Fix:* Check RepeaterBook. Band plans usually dictate this: If the frequency is below 147.000, it's usually -. If above, it's +. (But always verify!).

### 3. The "Alligator" (All Mouth, No Ears)

- *Symptom:* You can hear the repeater perfectly, it sounds strong. But when you talk, they say you are "in the mud" or unreadable.

- *Cause:* The repeater is a 100-watt monster on a 200-foot tower. It can "shout" to you easily. You are a 5-watt handheld in a valley. You can hear the shout, but you can't shout back loud enough to be heard.

- *Fix:* This is a physics problem, not a settings problem. Get a better antenna (Chapter 5), get outside of your car, or get to higher ground.

## ACTION STEPS

**Repeater Bring-Up Checklist**

- **Data:** Verified Output Freq, Offset, Direction, and Tone on Repeater-Book (Checked "Last Update" date to ensure it's not abandoned).

- **Program:** Entered into CHIRP or Keypad correctly.

- **Reverse Check:** Tapped * (Scan) button on the radio face to watch the frequency shift numbers.

- **Kerchunk:** Tapped PTT for 1 second. Heard the "tail" (static/beep) return.

- **Voice:** Made a radio check. Received a readable report from a human.

**"First Successful Repeater Contact" Script**

*Use this today. Don't wait for an emergency to feel awkward on the mic. Get a win.*

**You:** "Monitoring the 'Repeater Name' machine. This is "Your Callsign" *(Wait 10 seconds. This is polite. It lets people know you are there without demanding a response).*

**You:** "This is 'Your Callsign', looking for a quick signal report. I'm testing a UV-5R from "Your Location."

**Them:** "Your Callsign," this is "Their Callsign". You're loud and clear into the repeater."

**You:** "Thanks for the report, 'Their Callsign.' Just getting my settings dialed in. 'Your Callsign' clear."

# Chapter 4

# CHIRP 2.0 - THE BULLETPROOF PROGRAMMING WORKFLOW

IF PROGRAMMING A BAOFENG via the keypad is like writing a novel via text message on a flip phone, using CHIRP is like sitting down at a desktop workstation with a full keyboard and a mouse.

Keypad programming is a survival skill; CHIRP programming is a lifestyle skill. You need the survival skill for when the world falls apart, but you need the lifestyle skill to prepare for it.

CHIRP is free, open-source software that allows you to interface with your radio, manage hundreds of channels in a spreadsheet view, and unlock settings that are physically impossible to change from the radio's faceplate. It turns a tedious three-hour chore into a five-minute drag-and-drop operation.

However, CHIRP is also the place where most "Tactical Dads" quit. They buy the $20 cable, plug it in, get a Windows error message, fight with it for three hours, and then throw the cable in a drawer, never to be seen again. They assume they aren't "tech-savvy" enough.

This chapter is your technical support hotline. We are going to navigate the "Driver Hell" (and explain why it happens so you don't panic), fix the squelch settings that the factory broke, and build a channel management system that can adapt to hurricanes, road trips, and daily life without breaking a sweat.

# SECTION 1: THE GAUNTLET (CABLE & DRIVER SANITY CHECKS)

Let's address the elephant in the room immediately. You plugged the cable in, and it didn't work. You aren't crazy, and your computer isn't broken. You are a casualty of a corporate war.

## The "Counterfeit Chip" Conspiracy

Here is the reality: 99% of the cheap programming cables sold on Amazon (the $10 specials) use a cloned (counterfeit) "Prolific" chip. These are unauthorized copies of a chip designed by a company called Prolific.

Prolific got tired of losing money to cloners, so they did something aggressive: they updated their official Windows drivers to intentionally sabotage counterfeit chips. When Windows automatically updates your drivers (which it does constantly in the background), it grabs the latest official driver. That driver asks the chip, "Are you real?" The chip fails the test, and the driver shuts it down.

The result? Your computer sees the cable, but it displays a yellow warning triangle in Device Manager and refuses to talk to it. It might say "Code 10: Device cannot start."

**The Fix is NOT buying a new cable.** If you buy another $10 cable, you will have the same problem. The fix is telling Windows to ignore the new, aggressive driver and use the older, dumber driver (from 2008) that doesn't care about counterfeit chips.

## The "5-Minute Fix" (Step-by Step for Windows)

*Do not skip a step. Do exactly this. If you are on a Mac, you likely won't have this specific issue, but Windows users suffer this 90% of the time.*

1. **Plug it In:** Connect your programming cable to the USB port. Do NOT connect the radio yet.

2. **Open Device Manager:** Right-click the Windows Start button and select **Device Manager**.

3. **Find the Error:** Look for the section labeled "Ports (COM & LPT)". Click the arrow to expand it. You will likely see "Prolific USB-to-Serial Comm Port" with a yellow warning triangle icon next to it. That triangle is the sabotage in action.

4. **The Rollback:**

   - Right-click the device. Select **Update driver**.

   - Select **Browse my computer for drivers** (Do not search automatically).

   - Select **Let me pick from a list of available drivers on my computer**.

5. **The Magic Step:** You will see a list of driver versions.

   - The "Bad" Driver: Usually version 3.8.x or newer (dated 2015, 2019, etc.). This is the one stopping you.

   - The "Good" Driver: Look for version **3.2.0.0** (dated **2008** or **2007**).

   - *Troubleshooting:* If you do *not* see the 2008 driver in this list, Windows has deleted it. You need to go to a reputable site like Miklor.com, download the "Prolific 3.2.0.0 Installer," run it, and then restart this step.

6. **Select & Install:** Click the 3.2.0.0 version and click Next. Windows will install it instantly.

7. **Verification:** The yellow triangle should disappear immediately. Note the **COM Port Number** (e.g., COM3, COM4). You will need this number to tell CHIRP where to look.

## The Connection Ritual (Avoiding the "Radio Not Found" Error)

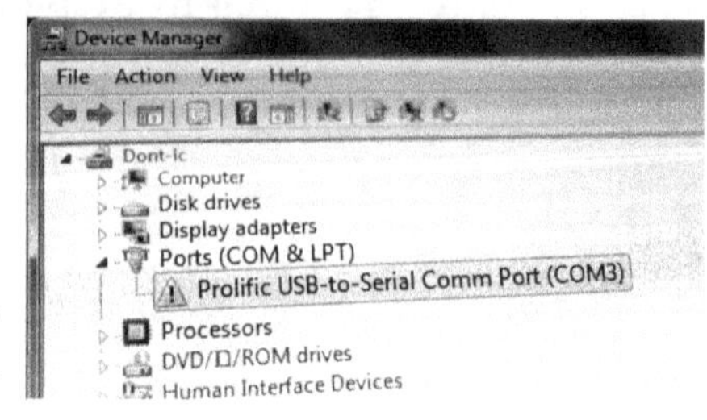


Once the driver is fixed, you must follow the ritual physically. The Baofeng connector (the two-prong K-plug) is notoriously tight on new radios. The plastic casing often prevents the pins from seating fully.

1. **Cable First:** Plug the USB end into the computer.

2. **Radio OFF:** Turn the radio volume knob OFF. (Connecting a live radio can sometimes trigger the Push-to-Talk).

3. **Plug In:** Insert the two-prong connector into the radio. **PUSH HARD.**

   - *The "Zero Gap" Rule:* You might feel one click. That is not enough. Push harder until there is **zero gap** between the cable head and the radio body. If you see metal pins, it's not in.

   - *Diagnosis:* If CHIRP says "Radio Not Found" or "Failed to communicate," 9 times out of 10, you simply need to push the plug in harder.

4. **Radio ON:** Turn the volume knob to 100%. A fully powered radio ensures the data signal is strong.

5. **Open CHIRP:** Go to Radio -> Download from Radio.

# SECTION 2: THE GOLDEN RULE OF PROGRAMMING

You are now connected. The green light is flashing on your radio. Before you type a single frequency, you must memorize the Golden Rule.

**READ > SAVE > EDIT > WRITE**

## 1. READ (Download from Radio)

Always start by pulling the current data *from* the radio. Even if the radio is brand new and empty.

   - *Why?* This "handshake" tells CHIRP exactly what firmware version your radio is running (e.g., N5R-20, BFB-297). CHIRP needs to know the map of the radio's brain before it can write to it.

## 2. SAVE (The "Factory Image")

The moment the download finishes, a tab will open showing your channels. **STOP.** Go to File -> Save As. Name it Baofeng_Factory_Backup_DATE.img

   - **Do not touch this file again.** Save it to your Dropbox, Google Drive, or a dedicated folder.

- *The "Soft Brick" Savior:* If you ever mess up your settings so badly the radio stops working, or you mess up the squelch settings and can't hear anything, this file is your time machine. Re-uploading this file returns the radio to the exact state it was in when you bought it.

## 3. EDIT

Now, make your changes. Delete the factory junk channels. Add your local repeaters. Change your settings. This is where the work happens.

## 4. WRITE (Upload to Radio)

Send the data back to the device.

### The "Firmware Rot" Warning

Do not take an image file you saved for a radio you bought in 2020 and try to force-upload it to a radio you bought in 2025. Even if they look identical and are both "UV-5Rs," Baofeng changes the internal memory locations constantly. If you force an old memory map onto a new chip, you can "soft brick" the radio (it lights up but screen is blank, or keys don't work).

- *The Safe Workflow:* Always download from the *new* radio to open a fresh tab. Then, open your *old* file, copy the channels (Ctrl+C), and paste them (Ctrl+V) into the new radio's tab. This moves the *data* without overwriting the *firmware map.*

# SECTION 3: TEMPLATES & MASS EDITING (YOUR LIFE RAFT)

Looking at a blank CHIRP spreadsheet is intimidating. Let's fill it using the **Channel Plan** we designed in Chapter 2.

### The Columns That Matter

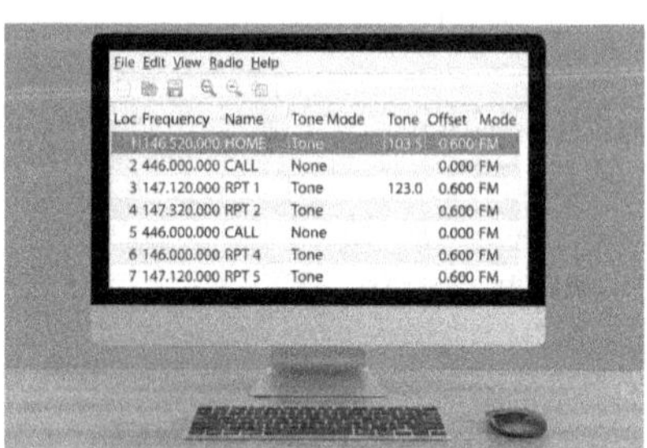

Ignore the columns that don't matter. CHIRP exposes everything, but you only need these:

- **Loc:** The channel number (0-127). This matches your "Virtual Zones."

- **Frequency:** The "Listen" frequency (Output).

- **Name:** 7 Characters max. Use ALL CAPS. (e.g., HOME-S, FIRE-L).

- **Tone Mode:** (See Chapter 3). Use Tone for Repeaters (Encode only). Use (None) for Simplex.

- **Tone:** The specific Hz value (e.g., 103.5).

- **Duplex:** + (Plus), - (Minus), or (None) for Simplex.

- **Offset:** The math (0.600, 5.000).

- **Mode:** Always NFM (Narrow FM) for Ham/GMRS/FRS. Using "FM" (Wide) is technically illegal on these frequencies now and will make you sound quiet to others.

- **Power:** High or Low.

- **Skip: Crucial.** If you check this box ("S"), the Scan function will skip over this channel.

## Importing External Data (The Fast Way)

Don't type repeaters manually. You will make a typo. CHIRP links directly to databases.

1. **Weather:** Go to Radio -> Import from Stock Config -> US NOAA Weather Alerts. This loads all 7 NOAA channels.

2. **Repeaters:** Go to Radio -> Query Source -> RepeaterBook.

3. **Filter:** Select your State and County. Select "2 Meters" or "70 Centimeters".

4. **Execute:** Click OK.

**The "Import" Window Strategy:** CHIRP will show you a list of 50 found repeaters. Do **not** just click OK. That clutters your radio with junk.

- **Uncheck All.** Start clean.

- Manually check the 3-5 repeaters you actually identified in your Chapter 2 plan.

- **Adjust the "To" Column:** This tells CHIRP where to put them. If you want them in your "Repeater Zone" (starting at Channel 20), change the first channel's "To" value to 20. The rest will automatically renumber (21, 22, 23). This keeps your memory map clean.

### Mass Editing (The Power Move)

This is the single best feature of CHIRP. Let's say you imported 20 channels, but they are all set to "Wide" FM and "Low" power, and you want them all "Narrow" and "High".

1. Click the first channel number row (e.g., Row 20).

2. Hold Shift and click the last channel number row (e.g., Row 40). The whole block is highlighted.

3. **Right-Click** anywhere in the highlighted block -> **Properties**.

4. A "Bulk Edit" window pops up.

5. Check the box for **Power** and set it to High.

6. Check the box for **Mode** and set it to NFM.

7. Click **OK**. *Boom.* You just updated 20 channels in 2 seconds. No more menu diving.

## SECTION 4: POWER USER TUNING (FIXING THE RECEIVER)

This is the secret sauce. This is the difference between a $25 toy and a usable tool.

The Baofeng's "Squelch" menu (SQL 0-9) is logically broken by default. Squelch is supposed to be a gate: Level 1 should be a flimsy gate (lets weak signals in), and Level 9 should be a bank vault (only strong signals get in). On factory Baofengs, Level 1 and Level 9 are often mathematically almost identical. This means you can't filter out interference without going completely deaf.

We can fix this in the **Service Settings**.

**The Squelch Threshold Fix**

1. In CHIRP, on the left sidebar, click **Settings**.

2. Select **Service Settings**. (If you don't see this, ensure you have "Developer Functions" enabled in View, though most modern builds show it by default).

3. You will see values for VHF Squelch and UHF Squelch. They look like weird numbers (e.g., 12, 15, 20...).

**The Logic:** These numbers represent the raw signal voltage required to open the squelch gate. We need to spread them out so 1 is sensitive and 9 is hard.

**Recommended "Tactical Dad" Values (Enter these):**

| SQL Level | VHF Value | UHF Value | What it feels like |
|:---:|:---:|:---:|:---:|
| 0 | 0 | 0 | Open (Constant Static) |
| 1 | 20 | 20 | Super Sensitive (Hear grass growing) |
| 2 | 25 | 25 | Sensitive (Good for rural areas) |
| 3 | 30 | 30 | **Standard Daily Use** |
| 4 | 40 | 40 | City Driving (Filters light noise) |
| 5 | 50 | 50 | Noisy Environment |
| 6 | 70 | 70 | High Interference (Near PC/Server) |
| 7 | 90 | 90 | Strong Signals Only |
| 8 | 110 | 110 | Very Strong |
| 9 | 125 | 125 | Next-to-Tower Only |

*Note: These are safe baselines. If you find Level 3 is still letting in too much static, bump the numbers up by 5.*

# SECTION 5: VERSIONING & SCENARIOS (THE NARRATIVES)

You don't wear a tuxedo to the gym. Why would you carry a "Disney World" channel plan during a "Hurricane"? Programming is not a "do it once" event. You should save different **Images** (.img files) for different missions.

**Narrative 1: The "Hurricane Prep" Image (Static Defense)**

**The Context:** A Category 3 storm is 48 hours out. You are staying put. Power will likely fail. Cell towers will likely jam or lose battery backups after 24 hours. **The Goal:** Conservation and Local Intel.

**The Setup Strategy:**

- **Channel 0 (Call):** 146.520 (National Simplex).

- **Channel 1 (Home):** Your family simplex channel. Crucial tweak: Set Power to **LOW** (1 Watt).

  - *Why?* You are all in the same house or neighborhood. You don't need 5 Watts to talk to the kitchen. Low power doubles your battery life.

- **Scan List:** VERY short. Home, ARES (Emergency) Repeater, and maybe the Fire Department dispatch. Do not scan 50 channels; you want to save battery cycles.

- **Weather:** NOAA channel programmed in memory but **removed** from scan. You don't want the radio stopping on weather constantly. You will manually switch to it at the top of the hour to check the track, then switch back.

- **Squelch:** Set to 1 or 2. You want to hear weak signals from neighbors asking for help, even if it means tolerating some static.

## Narrative 2: The "Disney Road Trip" Image (Dynamic Convoy)

**The Context:** You are driving two cars from Ohio to Florida. You have handhelds in both cars. **The Goal:** Convoy coordination and situational awareness on the highway.

**The Setup Strategy:**

- **Channel 1 (Convoy):** A GMRS or Ham simplex channel. Power set to **HIGH**.

  - *Why?* You are dealing with 70mph separation distances and metal car bodies blocking signals. You need the punch.

- **Tone Protection:** Use a DCS/CTCSS tone on your Convoy channel! The highway is RF soup. You don't want your wife to hear every random burst of static as you pass a Walmart or a truck stop. Lock the door with a tone.

- **Route Repeaters:** Use RepeaterBook's "Travel Search" feature. Import major repeaters for cities along I-75 (Cincinnati, Knoxville, Atlanta).

  - *Why?* You aren't chatting; you are listening for traffic accidents or road blockages reported by locals *before* you get stuck in the jam.

- **Scan List:** Convoy Channel (Priority) + Highway Patrol (if analog/legal) + MURS 3 (Truckers often use 151.xxx or CB, but MURS is common for road crews).

# SECTION 6: WORKBOOK TOOLS

## The CHIRP Workflow Card

*(Print this out, laminate it with packing tape, and tape it to your laptop lid or radio case)*

1. **Driver Check:** Device Manager -> Ports -> No Yellow Triangle? -> Good.

2. **Physical Connection:** Plug USB -> Radio OFF -> Plug K1 (HARD SNAP) -> Radio ON (100% Vol).

3. **Download:** Radio -> Download -> Select Port/Model -> OK.

4. **The Safety Save:** File -> Save As -> Backup_DATE.img.

5. **Edit/Import:**

   - Import from RepeaterBook (Uncheck all -> Select Top 3).

   - Mass Edit (Right-Click -> Properties) for NFM/High Power.

   - Organize Zones (0-19 Local, 20-59 Repeaters).

6. **Squelch Check:** Settings -> Service Settings -> Verify 20-125 spread.

7. **Upload:** Radio -> Upload to Radio.

8. **The Field Test:** Unplug. Turn off. Turn on. Press VFO/MR to go to Channel Mode. Check Channel 1.

## The "Programming Backup Plan" Checklist

*When the laptop dies, do you have a path?*

- **USB Drive in the Go-Bag:** Contains the "Prolific 3.2.0.0" installer .exe and the CHIRP installer .exe. (You can't download these if the internet is down).

- **Paper Repeater Map:** A printed list of the frequencies in your "Travel Image." If the radio resets, you need to type them in via keypad.

- **The "Rescue Cable":** A spare programming cable. They are flimsy. They break. Two is one, one is none.

Use this QR Code to Access Downloadable and Printable Materials Including CHIRP Template

# Chapter 5

# RANGE REALITY LAB

## ANTENNAS, PLACEMENT, AND PREDICTABLE PERFORMANCE

YOU BOUGHT THE RADIO. You looked at the box. It said, in bold, aggressive font: **"UP TO 36 MILE RANGE."**

You stood in your driveway. Your buddy stood at his house three miles away. You pressed the button.

Nothing.

You drove closer. Two miles. Nothing. You drove to the edge of his subdivision. One mile. Finally, you heard a scratchy, static-filled voice that sounded like a robot drowning in a blender.

You felt cheated. You assumed the radio was broken, or maybe you just bought a cheap piece of junk. You probably went online and saw people arguing about "fake chips" or "bad batches."

Here is the truth: The radio is fine. The box is a lie. And physics is a harsh, unforgiving mistress.

This chapter is the "Red Pill" of radio. We are going to strip away the marketing hype and look at the cold, hard math of Radio Frequency (RF). By the end of this chapter, you won't just know *why* you couldn't talk three miles; you will know exactly how to fix it so you can. We are going to turn "magic" into "mechanics."

# SECTION 1: WHY RANGE CLAIMS LIE (THE PHYSICS OF LINE OF SIGHT)

How can they legally print "36 Miles" on the box?

The manufacturer's claim assumes a specific, impossible scenario: You are standing on the peak of a mountain. Your friend is standing on the peak of another mountain 36 miles away. There is a perfect vacuum between you. There are no trees, no buildings, no power lines, and no atmosphere to scatter the signal. In that specific, theoretical universe, yes, 5 watts will travel 36 miles.

In the real world, you are standing in a valley (your driveway), surrounded by RF-absorbing sponges (pine trees), inside a Faraday cage (your car), trying to push a signal through six blocks of stucco, copper wiring, and rebar.

## The "Line of Sight" Rule

VHF (Very High Frequency) and UHF (Ultra High Frequency), the bands your Baofeng uses, are **Line of Sight** waves. They behave almost exactly like light from a flashlight.

If you shine a flashlight at your neighbor's house, but there is a hill in the way, the light hits the dirt. It does not curve over the hill. It does not go through the hill. It stops.

Radio waves are the same. If there is dirt, rock, or a dense cluster of buildings between your antenna and his antenna, the signal dies. It doesn't matter if you have 5 watts or 50 watts; you cannot power your way through a mountain.

**The Equation of "Height is Might"** This is the only rule that matters. Power (Watts) is overrated. Height is everything.

- **5 Watts at 5 feet off the ground:** Range ~1-2 miles in suburbs. You are fighting bushes, cars, and fences.

- **5 Watts at 50 feet off the ground:** Range ~10-15 miles. You have cleared the "clutter" of the neighborhood.

- **5 Watts on a mountain top:** Range ~50+ miles. You have a direct line of sight to the horizon.

*Tactical Dad Lesson:* If you can't reach someone, don't shout louder (add power).

Climb higher (add height). Standing on the bed of your truck, climbing a deer stand, or walking to the second floor of your house does more for your signal than buying a $200 amplifier.

## The "Sponge" Factor (Attenuation)

Radio waves pass through some things and get absorbed by others. This is called **Attenuation**. Think of it like running through water.

- **Glass/Drywall:** Easy. Minimal loss. You can talk through a wood-frame house easily.

- **Leaves/Trees:** Medium to Heavy loss. Pine needles are specifically tuned by nature to be approximately the length of 70cm waves. A dense forest is basically a "radio black hole" for UHF signals. Wet leaves after a rainstorm are even worse.

- **Concrete/Rebar:** Heavy loss. Urban environments are tough because every building is a cage of steel mesh.

- **Earth/Hills:** Total blockage. Game over. There is no "punching through" the earth.

## The Urban Canyon Effect (Multipath)

In a city, you might find that you can't talk to someone three blocks away because of buildings, but you *can* talk to someone if you turn 90 degrees and bounce the signal off a glass skyscraper.

This is **Multipath**. Your signal bounces like a billiard ball. Sometimes these bounces help you (reflecting around a corner), and sometimes they hurt you (two signals arriving at different times, canceling each other out). If you are in a city and the signal is garbage, move five feet to the left. You might step out of a "dead spot" and into a reflection sweet spot.

## SECTION 2: ANTENNA UPGRADES THAT ACTUALLY HELP

Your radio came with a short, stiff, rubber-coated antenna. We call this the "Rubber Ducky." It is durable, small, and convenient.

It is also an electrical compromise. It is a "dummy load on a stick." To make it 4 inches long, the manufacturer coiled the wire tightly, which creates resistance and heat. A huge chunk of your battery power is wasted just heating up the rubber coating, rather than radiating out as a signal.

## Upgrade 1: The "Whip" (Nagoya NA-771 / Signal Stick)

The single best $20 you can spend is replacing the stock antenna with a "tuned" 1/4 wave or 5/8 wave whip. These are usually 15 to 19 inches long.

- **The Difference:** A stock antenna might radiate 30% of your power. A 15-inch whip might radiate 70-80%. That is effectively doubling your output range without draining your battery faster.

- **Recommended Brands:**

  - **Nagoya NA-771:** The classic. It has a rigid base and a whippy top. *Warning:* There are more fake Nagoyas on Amazon than real ones. Buy from a reputable dealer (like BTech).

  - **Signal Stuff "Signal Stick":** Made in the USA by ham radio nerds (HamStudy.org). It is made of **Nitinol** (memory alloy). You can tie it in a literal knot, stuff it in your pocket, sit on it, and when you pull it out, it snaps straight instantly. It is lighter and more durable than the Nagoya. **This is the Tactical Dad choice.**

## Upgrade 2: The "Tactical Foldable" (The Tape Measure)

You will see antennas that look like folding rulers or flexible steel tape measures (like the Abbree AR-771). They look very "commando."

- **The Truth:** They are electrically decent, but they are physically heavy.

- **The Risk:** The heavy base puts massive leverage stress on the radio's SMA connector. If you drop the radio, or even just wave it around too vigorously, the leverage can snap the connector right off the circuit board.

- **The Verdict:** Use them if you need to fold the radio into a pouch for transport. But when you deploy it, hold the base of the antenna with your thumb to support the weight.

## Upgrade 3: The Vehicle Mag Mount (The Game Changer)

This is where the magic happens. If you use a handheld radio inside a car, you are trapped in a metal box (Faraday Cage). The metal body blocks the signal from getting out, and the electronic noise from the engine (alternator whine) interferes with the signal getting in. You will get maybe 0.5 miles of range.

**The Fix:** A "Mag Mount" (Magnetic Mount) antenna placed on the **center of the roof**.

**Why it works:** It's not just that the antenna is outside. The steel roof of your car acts as a **Ground Plane**. It becomes the "other half" of the antenna system, reflecting the signal upward and outward like a mirror.

- **The Result:** A $25 mag mount antenna (like a Nagoya UT-72) connected to your $25 Baofeng will outperform a $500 handheld sitting in the cup holder. **Range often jumps from 0.5 miles to 5-10 miles instantly.**

### The Secret Weapon: The "Tiger Tail" (Counterpoise)

Every antenna needs two halves: the radiating element (the stick) and the ground (the other side). On a car, the roof is the ground. On a handheld, **YOU** are the ground. The radio uses your hand and arm to complete the circuit. This is inefficient.

- **The Hack:** A "Tiger Tail" or "Rat Tail" is simply a piece of wire crimped to a ring terminal that slips over the antenna connector and hangs down freely.

- **Length Matters:**

  - **Ham Radio (2m):** ~19 inches.

  - **GMRS / 70cm:** ~6.5 inches.

- **The Effect:** It provides a proper electrical ground. It creates a "Dipole" antenna.

- **The Value:** It costs about 50 cents to make and can boost your signal by 30-40% in marginal conditions.

# SECTION 3: PLACEMENT RULES (BODY LOSS & GEOMETRY)

Where you hold the radio matters more than the radio itself. You can have the best antenna in the world, but if you bury it in your armpit, it won't work.

## The "Meat Sack" Problem (Body Blocking)

You are mostly a bag of salt water. Salt water is excellent at absorbing radio waves.

- **The Scenario:** You clip the radio to your belt. You wear a headset. The antenna is pressed tight against your kidney.

- **The Physics:** Your body acts as a massive shield. You are absorbing 60-80% of the signal radiating to your rear and sides. You are "shadowing" the signal. You can hear people in front of you, but you are deaf to people behind you.

- **The Fix:** Get the antenna away from your body.

  - *Good:* Holding it in your hand, arm extended slightly, radio vertical.

  - *Better:* Radio in a chest pouch or high on a backpack strap where the antenna clears the shoulder.

  - *Best:* External antenna up a tree or on a mast.

## Polarization (Vertical vs. Horizontal)

Baofeng antennas are "Vertical." They transmit waves that ripple up and down. They want to stand straight up.

- **The Mistake:** You hold the radio sideways "gangster style" or lay it flat on a picnic table while eating lunch.

- **The Penalty:** "Cross-Polarization Loss." If your antenna is horizontal and the repeater antenna (or your buddy's antenna) is vertical, you lose about **20dB** of signal.

    ◦ *Translation:* You arc rccciving 1/100th of the signal strength. That is the difference between "Loud and Clear" and "Total Silence."

- **The Rule:** Keep the antenna vertical. Even if you are lying prone, tilt the radio so the antenna points to the sky.

## SECTION 4: COAX AND CONNECTORS (AVOIDING THE VAMPIRES)

If you decide to connect your radio to an external antenna (like on your roof or in your attic), you need a cable. This cable is called "Coax" (Coaxial Cable).

### The Gender Confusion (SMA-Female vs. Male)

Baofeng defied industry standards just to mess with us.

- **Most Handhelds (Yaesu/Icom):** The radio has a Pin (Male). The antenna has a Hole (Female).

- **Baofeng:** The radio has a **Pin** (SMA-Male). The antenna has a **Hole** (SMA-Female).

**The Trap:** You buy an antenna meant for a different brand (like Wouxun or Anytone) and try to screw it on. It fits physically, the threads match, but **neither side has a pin.** Or worse, **both sides have pins. The Consequence:** You transmit into an open circuit. The energy reflects back into the radio and fries the final transistor. Your radio is now a brick. **The Fix:** Always verify the center pin. Look before you screw. One side must have a pin; the other must have a hole.

**Adapters are Signal Vampires**

Every time you convert from SMA to BNC to UHF to N-Type, you lose signal. Each adapter "eats" a fraction of a decibel.

- *Bad:* Baofeng -> SMA-to-BNC adapter -> BNC-to-UHF adapter -> Cable.

- *Good:* Baofeng -> Cable with SMA connector pre-installed.

- *Rule:* Use the fewest adapters possible.

**Coax Loss (The Garden Hose Analogy)**

Think of Coax like a garden hose. Signal is water pressure. Friction kills pressure.

- **RG-58 (The Thin Stuff):** This is like a drinking straw. It's cheap and flexible.

  - *The Math:* If you run 50 feet of RG-58 to a UHF antenna, you lose half your power just in the cable.

  - *Use Case:* Fine for short mobile runs (10-15 ft) in a car. Terrible for base stations.

- **RG-8x (The Medium Stuff):** A standard garden hose. Good for 50ft runs. A great balance of flexibility and performance.

- **LMR-400 (The Thick Stuff):** A fire hose. It is thick, stiff, and expensive. But it loses almost no signal.

- **Tactical Dad Rule:** For a car mag mount, the thin cable attached to it is fine. Route it through the **top corner** of the door frame where the weather stripping is thickest/softest. Never pinch coax in a car door jamb or window, if you crush the foam inside, the cable is ruined.


# SECTION 5: RUN A RANGE TEST LIKE A GROWN-UP

Stop asking "Can you hear me?" and start collecting data. You need to map your "Circle of Influence."

Knowing you have a "2-mile hard deck" is better than *hoping* you have a 10-mile range.

**The "Tactical Dad" Range Protocol**

You need two people. One at "Base" (Home) and one "Rover" (Mobile).

**1. The Setup:**

- **Base Station:** Wife/Kid/Buddy at home. Radio set to **High Power**. Antenna vertical (standing up on the counter, or preferably upstairs near a window).

- **Rover:** You in the car (or on a bike). Radio set to **High Power**.

**2. The Route:** Drive in a cardinal cross (North, South, East, West). Pick landmarks at 0.5 mile, 1 mile, 2 miles, and 5 miles.

**3. The Check-In Script:** Don't just say "Hello." Use the **RST** format (Readability, Signal).

- **Readability:** 1 (Unreadable) to 5 (Perfect).

- **Signal:** 1 (Static noise) to 9 (Strong/Quiet).

- *Rover:* "Base, this is Rover. Reaching Checkpoint Alpha. 1 Mile North. How copy?"

- *Base:* "Rover, this is Base. I copy you 5 by 9. Loud and clear. How me?"

- *Rover:* "Base, I copy 5 by 9. Proceeding to Bravo."

**4. The Failure Point (The "Fade"):** Eventually, you will hit a spot where Base says: *"Rover, I copy you... 3 by... static... broken."* **Stop.** Do not give up yet.

- **Audit the Environment:** Are you in a valley? Are there power lines?

- **The Car Pivot:** Turn the car 90 degrees. The metal frame of the car might be blocking the path back home.

- **The Jailbreak:** Get out of the car. Stand up. Hold the radio high. Did the signal improve? (Usually yes, dramatically).

**5. Logging the Data:** Go home and open Google Maps (or a paper map). Drop pins where you lost signal. Connect the dots. **This is your Real World Range.**

It doesn't matter that the box said 36 miles. Your map says 2.4 miles to the North and 0.8 miles to the South because of that hill. *This map is the most valuable survival tool you own. It tells you exactly where you need to drive to send a message.*

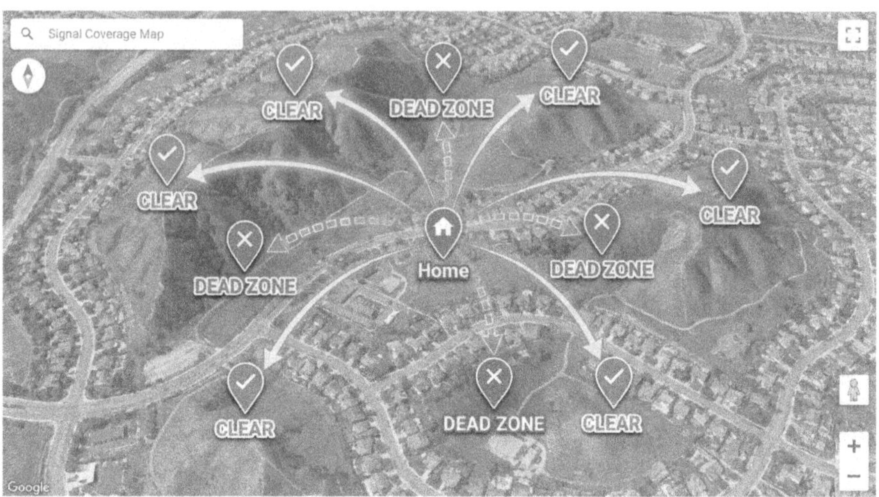


## ACTION STEPS & WORKBOOK

### Range Test Log

*Print this and take it in the car.*

- **Date:** _______________

- **Weather:** _________________________

  - (Rain kills UHF range! Test in bad weather to know your "worst case.")

- **Base Antenna:** _________________

- **Rover Antenna:** _______________

| Distance | Direction | Location/Landmark | Readability (1-5) | Notes (Static? Choppy? Had to get out of car?) |
| --- | --- | --- | --- | --- |
| 0.5 Mile | North | *Example:* Gas Station | 5 | Perfect audio. |
| 1.0 Mile | North | | | |
| 2.0 Miles | North | | | |
| 3.0 Miles | North | | | |
| Limit | North | *Example* | 1 | Lost signal at the creek bridge. |
| 1.0 Mile | South | | | |
| 2.0 Miles | South | | | |
| Limit | South | | | |
| 1.0 Mile | East | | | |
| Limit | East | | | |

## The "Placement Rules" Cheat Sheet

*Tape this to your radio storage bin.*

1. **Height is Might:** 1 foot of height = 1 mile of range (roughly). Get high.

2. **Vertical Always:** Never transmit with the radio laying flat.

3. **Escape the Cage:** Inside a car = 90% signal loss. Get an external antenna or get out.

4. **Body Block:** Keep the antenna away from your ribs. High on the shoulder is best. The radio needs to "see" the horizon.

5. **The "Baofeng Twist":** Ensure your adapter has a HOLE (SMA-Female) to match the radio's PIN (SMA-Male). If you see two holes, you are killing your radio.

# Chapter 6

# AUDIO DISCIPLINE & ON-AIR CONFIDENCE

YOU HAVE THE RADIO. You programmed the frequencies. You climbed a hill to get a signal. You press the button to speak.

*"Uhh, hey honey, it's me. Are you there? Uh... 10-4? Over?"*

You sound terrified. You sound unsure. And worse, because you held the microphone wrong, you sound like you are speaking from inside a running blender.

This is the most overlooked skill in radio. We spend hours obsessing over antennas, watts, and batteries, but we spend zero minutes thinking about **cadence** and **clarity**. We treat the radio like a cell phone, assuming the technology will fix our bad habits. It won't.

**Here is the brutal reality:** If your signal is strong (5 bars) but your audio is garbage, **you have failed.** If your audio is crisp but your message is rambling and confusing, **you have failed.**

In a high-stress situation, when the power is out, the storm is raging, or you are separated from your group, your voice is the only thing your family has to hold onto. If you sound panicked, high-pitched, and breathless, they will panic. If you sound calm, clear, and professional, they will calm down. Your voice is a tool for command and control.

This chapter teaches you how to sound like a "Level 4 Operator" even if you are just asking if we need milk. We are going to cure "Mic Fright" and replace it with muscle memory.

# SECTION 1: HOW TO SOUND CLEAR ON A CHEAP HT

The microphone on a $25 Baofeng is an electret condenser element about the size of a grain of rice. It is not a studio microphone. It has zero pop-filtering, terrible wind rejection, and very poor dynamic range. If you scream into it, it doesn't get louder; it just turns your voice into a square-wave of distortion. You have to work *with* the hardware, not against it.

## 1. The "Baofeng Delay" (The First Syllable Killer)

This is the single most common mistake new users make, and it makes you sound like an amateur instantly.

Modern radios have "Battery Save" features. To extend life, the receiver rapidly turns itself on and off (sleep/wake) hundreds of times a second. When a signal comes in, it takes about 500 milliseconds (half a second) for the chip to say, "Oh, that's a signal," wake up the audio amplifier, and open the speaker.

Furthermore, repeaters are slow. When you press PTT (Push-to-Talk), your radio has to send a CTCSS tone to the repeater. The repeater has to "hear" the tone, verify it, and then turn on its own transmitter. This mechanical handshake takes time.

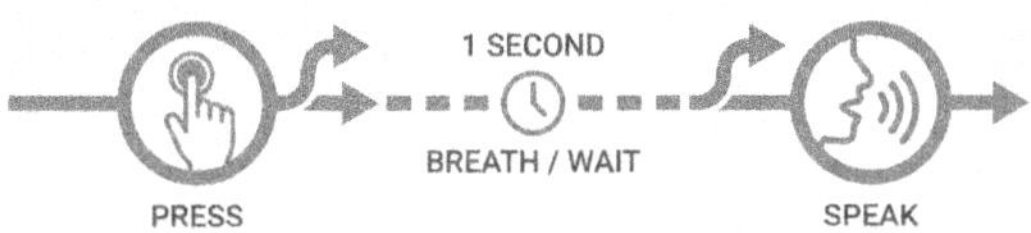


**The Mistake:** You press the button and speak instantly.

- *You say:* "Base, this is Rover."

- *They hear:* "...ver."

- *Result:* Confusion. "Say again? Who is this?"

**The Fix: The "Press... Breathe... Speak" Rhythm.**

1. **Press** the PTT button firmly.

2. **Take a short breath** (roughly 1 second). This feels like an eternity, but it is necessary.

3. **THEN** speak.

*Tactical Dad Tip:* Use this breath to calm your nerves. That 1-second pause guarantees the line is open, the repeater is active, and your family hears the *first* letter of your name.

## 2. The "Ice Cream Cone" Grip

Where you hold the radio changes the audio physics.

- **The "Pizza Slice" (Bad):** Holding the radio flat in front of your face like you are eating a slice of pizza. You are speaking *over* the mic, not *into* it. Your voice will sound distant, hollow, and echoey.

- **The "Rock Star" (Bad):** Eating the mic. Lips touching the plastic. You are blasting air directly into the diaphragm.

  - *The Physics:* Fast-moving air from "P" and "B" sounds (Plosives) hits the diaphragm like a hammer. This causes "clipping" (distortion). You will sound like muffled garbage or a blown-out speaker.

- **The "Ice Cream Cone" (Good):** Hold the radio vertical, about 2-3 inches away from your mouth, slightly to the side (45-degree angle).

  - *The Technique:* Speak **across** the mic hole, not directly **into** it. The sound waves enter the mic, but the burst of air from your mouth passes harmlessly by. This creates clear, distortion-free audio and helps cut down on wind noise when you are outdoors.

## 3. Volume and Cadence (The Adrenaline Dump)

When humans are stressed, two things happen biologically:

1. Our vocal cords tighten (Voice gets higher).

2. We speak faster (trying to dump information).

Radio circuits are designed for normal conversation volume in a lower register. If you shout, the audio "clips" (hits the ceiling of what the radio can process) and turns into jagged static. A whisper is equally bad because it gets buried in the noise floor.

- **Speak Slower:** Under stress, force yourself to talk at 75% of your normal speed. Enunciate every syllable.

- **Speak Monotone:** This sounds weird, but a calm, flat tone cuts through static better than an excited, high-pitched one. Think "Airline Pilot" encountering turbulence, not "YouTube Streamer" playing a video game.

  - *Bad:* "Ohmygod there's a tree down huge tree right across the road!"

  - *Good:* "Break. Tree down. Blocking road. Large size."

## SECTION 2: THE "ANTI-CB" MANIFESTO (PLAIN ENGLISH ONLY)

Somewhere along the way, pop culture convinced us that using radios meant talking like a trucker from a 1970s movie. *"Breaker breaker one-nine, what's your 10-20? We got a Smokey taking pictures."*

**Stop it.** Remove this from your brain immediately. Unless you are a trucker or a police officer in a specific jurisdiction, **do not use 10-codes.**

### Why 10-Codes are Dangerous

Codes are not universal. They are a "local dialect."

- "10-50" means "Traffic Accident" in one county.

- It means "Officer Down" in the next county.

- It means "Traffic Stop" in a third.

In a regional disaster (hurricane, wildfire), you might be talking to volunteers, GMRS users, Ham operators, and National Guard members from three different states. If you use a code, you risk being misunderstood. Confusion kills. Even the Federal Government (NIMS/ICS) has mandated "Plain English" for all inter-agency disaster communications.

**The "Tactical Dad" Standard: Plain English**

Say what you mean. Use standard words. Be boringly obvious.

- Instead of "What's your 10-20?", say **"What is your location?"**

- Instead of "10-4," say **"Received"** or **"Copy."**

- Instead of "10-9", say **"Say again."**

**The "Pro-Words" (Professional Words)**

While we avoid codes, we *do* use a specific set of standardized words (Pro-Words). These are not codes; they are **procedure**. They function like punctuation marks in a sentence, telling the other person exactly what is happening.

| The Word | What It Means | When To Use It |
| --- | --- | --- |
| "COPY" | I heard and understood your message. | The universal acknowledgment. |
| "WILCO" | "Will Comply." I heard you AND I will do it. | When given an instruction. (Don't say "Roger Wilco," it's redundant.) |
| "STANDBY" | I am busy. Wait. I will call you back. | When you need a minute to think, drive, or write something down. |
| "SAY AGAIN" | I didn't hear you. Repeat the message. | *Never say "Repeat."* (In military contexts, "Repeat" is a specific command to artillery to fire the same mission again. Don't call in artillery on your house.) |
| "AFFIRMATIVE" | Yes. | "Yes" is a short, soft word that gets lost in static. "Affirmative" is distinct. |
| "NEGATIVE" | No. | "No" is too short. "Negative" cuts through noise. |
| "BREAK" | I am separating two different messages. | "I need milk. BREAK. I also need gas." Use this instead of unkeying the mic. |
| "CORRECTION" | I made a mistake. Here is the right info. | "Turn Left... CORRECTION... Turn Right." |
| "OUT" | I am ending the conversation. | "Copy. Out." (Do not say "Over and Out." That means "I'm done talking, go ahead, but I'm hanging up.") |

# SECTION 3: THE LANGUAGE OF CONFIDENCE (NATO PHONETIC ALPHABET)

This is the mark of a pro. This is the difference between "Did you say B or D?" and instant clarity.

In a static-filled environment, the human ear struggles to distinguish rhyming letters. This is the **"Kill Zone" of the alphabet:**

- **B, C, D, E, G, P, T, V, Z.** They all end in "EEEE."

- **M and N.** They sound identical. **F and S.** They both sound like hiss.

To fix this, NATO developed a set of words where **no two words sound alike.** Even if you only hear half the word "November," you know it isn't "Mike."

**The Deep Dive: How to Learn It**

Do not just print the chart and stick it in your wallet. You have to wire this into your brain so you can speak it without thinking. When you are spelling your street name to emergency services over a dying battery, you don't want to be looking for a cheat sheet.

**The Chart (Tactical Dad Reference)**

| TACTICAL ALPHABET | | |
| --- | --- | --- |
| A - ALPHA | J - JULIETT | S - SIERRA |
| B - BRAVO | K - KILO | T - TANGO |
| C - CHARLIE | L - LIMA | U - UNIFORM |
| D - DELTA | M - MIKE | V - VICTOR |
| E - ECHO | N - NOVEMBER | W - WHISKEY |
| F - FOXTROT | O - OSCAR | X - X-RAY |
| G - GOLF | P - PAPA | Y - YANKEE |
| H - HOTEL | Q - QUEBEC | Z - ZULU |
| I - INDIA | R - ROMEO | |

**Number Discipline (The Forgotten Skill)**

Just like letters, numbers get confusing. "Nine" sounds like "Five" in static. "Nine" also sounds like "Nein" (No) in German, which caused issues in NATO.

- **9** is spoken as **"NINER"** (To distinguish from Five/None).

- **5** is spoken as **"FIFE"** (To distinguish from Fire).

- **4** is spoken as **"FOWER"** (To distinguish from "For" or noise).

- **3** is spoken as **"TREE"** (To keep it sharp).

- **0** is spoken as **"ZERO"** (Never say "Oh").

### The "License Plate" Drill

This is how you learn this in 3 days without flashcards. While driving to work, look at the license plate of the car ahead of you. Translate it instantly.

- Plate: K7P M29

- You say (out loud): *"Kilo Seven Papa... Mike Two Niner."*

- Plate: DVG 304

- You say: *"Delta Victor Golf... Tree Zero Fower."*

Do this for 10 cars a day. It forces your brain to recall the words randomly and rapidly. By Friday, you will be fluent.

## SECTION 4: THE "C.L.E.A.R." REPORTING FORMAT

When you get on the radio, you shouldn't ramble. Rambling ties up the frequency, drains the battery, and annoys everyone listening. *"Uhhh, hey, so I'm at the store, and it's kinda crazy, lots of people, um, I think they are out of water, but I saw some Gatorade, do you want that?"*

**No.** That transmission took 15 seconds to say nothing. We need a standard format. The military uses "SALUTE" reports. We are going to use a civilian adaptation designed for families.

### The C.L.E.A.R. Report

- **C - CALLSIGN:** Who are you? (Identify yourself first).

- **L - LOCATION:** Where are you? (Be specific).

- **E - EVENT:** What is happening? (The headline).

- **A - ACTION:** What are you doing about it? (Your status).

- **R - REQUEST:** What do you need? (Resources).

### Example 1: The Grocery Run (Low Stress)

*"Base, this is Dad. **Location:** Kroger on Main. **Event:** Store is crowded, out of water. **Action:** I am buying Gatorade instead. **Request:** Do we need bread? Over."*

**Example 2: The Car Breakdown (High Stress)**

*"Base, this is Unit 1. **Location:** I-75 Mile Marker 102. **Event:** Flat tire, safe on shoulder. **Action:** Changing tire now. **Request:** Stand by for check-in in 15 minutes. Over."*

**Why this works:** Even if your wife misses the first part because of static, the structure tells her what is coming. She knows the second thing you say is *Where* and the last thing you say is *What you need*. It forces you to think before you speak. It stops the panic-rambling.

# SECTION 5: ETIQUETTE & "DON'T BE THAT GUY"

Radio communities are small. If you act like a goofball, you will be blocked, ignored, or shamed.

**The "Don't Be That Guy" List**

1. **The "Roger Beeper":**

   - *The Crime:* Having that annoying "BEEP" sound at the end of every transmission. (Menu 39 ROGER on UV-5R).

   - *Why it's bad:* NASA used Quindar tones because they were talking to the moon. You are talking to the suburbs. On a cheap radio, this beep is often 2x louder than your voice, blasting the eardrums of anyone wearing a headset. It marks you as a newbie immediately.

   - *Fix:* Turn it OFF.

2. **The "Baofeng Siren" (The Orange Button of Shame):**

   - *The Crime:* Accidentally bumping the orange CALL button on the side and sending a loud, wailing siren alarm over the airwaves.

   - *Why it's bad:* It sounds like an emergency, but it's usually just you sitting on your radio. It deafens headphone users.

   - *Fix:* Be very careful with the orange button. On some models, you can disable the "Alarm" function in CHIRP (Set "Alarm Mode" to "Site" instead of "Tone").

### 3. The "Kerchunker":

- *The Crime:* Keying the mic repeatedly to hear the repeater tail ("Ker-chunk"), but never identifying yourself.

- *Why it's bad:* It triggers the repeater for everyone listening. It's like ringing a doorbell and running away.

- *Fix:* If you key up, say your callsign. "Testing, this is 'Callsign'."

### 4. The "Dead Key-er":

- *The Crime:* Sitting on the PTT button while you think of what to say.

- *Why it's bad:* You are blocking everyone else from using the channel. In radio, only one person can talk at a time. If you are transmitting "Uhhhh...", nobody else can call for help.

- *Fix:* **Think** *then* **Push**. Not Push *then* Think.

### 5. The "Interrupting Cow":

- *The Crime:* Jumping in the second someone stops talking.

- *Why it's bad:* In emergency nets, we leave a "gap" between transmissions to let urgent traffic break in.

- *Fix:* When you hear someone finish ("...clear"), count "One Mississippi" before you press your button. This gap allows a third party with an emergency to shout "Break!"

### 6. The "Heavy Breather" (Accidental VOX):

- *The Crime:* Turning on VOX (Voice Operated Switch) by mistake and broadcasting every heavy breath or gust of wind.

- *Why it's bad:* It keeps the channel open constantly, drains your battery, and annoys everyone.

- *Fix:* Ensure VOX is set to **OFF** (Menu 4) unless you are specifically setting it up for a controlled task.

# WORKBOOK TOOLS

## The "Message Discipline" Pocket Card

*Print, laminate, and zip-tie to your radio or backpack.*

### BEFORE YOU PRESS PTT:

1. **LISTEN:** Is the channel clear?

2. **THINK:** What is my C.L.E.A.R. message?

3. **WAIT:** Press PTT, wait 1 second (breath), then speak.

### THE C.L.E.A.R. FORMAT:

- **C**allsign (Me to You)

- **L**ocation (Where)

- **E**vent (What)

- **A**ction (Doing)

- **R**equest (Need)

### PRO-WORDS:

- **COPY:** Understood.

- **WILCO:** Will do it.

- **SAY AGAIN:** Repeat.

- **CORRECTION:** I messed up.

- **BREAK:** New topic / Interruption.

## 5 Practice Scenarios (The Drill)

*Do these with your partner. One person in the kitchen, one in the yard. Grade each other on Clarity, Phonetics, and Format.*

**Scenario 1: The Storm Spotter** *Context:* You see a tree down blocking the neighborhood entrance. *Goal:* Report it using C.L.E.A.R. and phonetic spelling of the street name. *Drill:* "Base, this is Unit 1. Location: Entrance to **OAK** subdivision. That is **OSCAR-ALPHA-KILO**. Event: Large tree blocking road. Action: Assessing if I can move it. Request: Call public works. Over."

**Scenario 2: The Lost Kid** *Context:* You are at the county fair. You can't find your son, Jason. *Goal:* Communicate urgency without panic. Use description. *Drill:* "Base, this is Dad. Location: Ferris Wheel. Event: Separated from Jason. Action: Starting grid search north. Request: You stay at the gate. Description: Red Shirt, Blue Hat. Over."

**Scenario 3: The Bad Connection** *Context:* The signal is scratchy. You need to verify an address (15 Maple St). *Goal:* Use "Say Again" and Number Discipline. *Drill:* "Base, signal is broken. Say again address. Over." -> "I say again: **ONE-FIFE** Maple. **MIKE-ALPHA-PAPA-LIMA-ECHO**. Over."

**Scenario 4: The Convoy** *Context:* Driving two cars. You need to stop for gas. *Goal:* Brevity. Don't clog the channel. *Drill:* "Unit 2, this is Unit 1. Fuel low. Exiting next ramp. Wilco? Over." -> "Unit 1, this is Unit 2. Copy. Exiting next ramp. Out."

**Scenario 5: The Radio Check** *Context:* Testing a new antenna. *Goal:* Correct format and patience. *Drill:* "Radio Check, Radio Check. This is 'Name' on 'Channel Name'. How copy? Over."

# Chapter 7

# POWER, CHARGING, AND ENDURANCE

## THE PART EVERYONE REGRETS IGNORING

You have the best radio. You have a tuned antenna. You have a perfect channel plan. You have practiced your phonetic alphabet until you dream in "Alpha Bravo."

Then, 12 hours into the blackout, you hear the dreaded *beep-beep*. The screen flashes. The radio dies.

You reach for your charger, but the power is out, so the wall outlet is useless. You run to your car, but you realize your charging cradle requires a 10V wall wart, not a USB plug. You stare at a pile of AA batteries in your junk drawer, but you have no way to put them into the radio.

Your communications plan just failed. Not because of a lack of skill, but because of a lack of electrons.

Most people treat batteries as an afterthought. They use the stock battery, charge it once a month, and assume it will be there when the world ends. This is a fatal error. In a real emergency, **power is the fuel of information.** If you run out of fuel, you are deaf and mute.

This chapter is about hardening your power supply. We are going to move away from proprietary, clunky chargers and embrace a standardized, redundant system that keeps you on the air for weeks, not hours.

# SECTION 1: BATTERY TRUTH (THE PHYSICS OF THE BRICK)

The standard battery that comes with a UV-5R (the BL-5) is a Lithium-Ion pack rated at roughly 1800mAh (milliamp-hours). Note that I said "rated at." Cheap Chinese batteries often lie.

That "3800mAh" extended battery you saw on eBay for $12? It's probably just an 1800mAh cell with a bigger plastic shell filled with hot glue and lead weights. Unless you buy from a reputable brand (like BTech or Baofeng Tech), assume the label is optimistic fiction.

**The "Duty Cycle" Reality**

To understand endurance, you have to understand consumption. Your radio burns power at three drastically different rates. Understanding this math is the difference between your radio lasting 6 hours or 3 days.

1. **Standby (The "Listening" Mode):**

   ○ *Draw:* ~75 milliamps.

   ○ *Life:* If you just leave the radio on and nobody talks, a healthy battery lasts about **24 hours**. This is your baseline.

2. **Receive (The "Talking" Mode):**

   ○ *Draw:* ~300-400 milliamps (depending on volume).

   ○ *Life:* If you are listening to a busy repeater conversation non-stop, or monitoring a weather alert that is broadcasting continuously, the battery lasts about **4-6 hours**. The audio amplifier consumes significant power just to push sound out of the speaker.

3. **Transmit (The "Flamethrower" Mode):**

   ○ *Draw:* ~1500 milliamps (1.5 Amps) on High Power (5W).

   ○ *Life:* If you held the PTT button down continuously, the battery would die in **less than 1 hour** (and the radio would likely melt or go into thermal shutdown). Transmitting is expensive.

**The "5-5-90" Rule:** Engineers design radios assuming a 5-5-90 duty cycle:

- 5% of the time Transmitting (Brief reports).

- 5% of the time Receiving (Listening to replies).

- 90% of the time on Standby (Silence).

- *Result:* This gives you a "shift life" of about 10-12 hours. If you talk more, you die faster.

**The Vampire Settings (What Kills You)**

If your battery is dying in 6 hours, you are probably guilty of these crimes. These are the "Vampires" that suck the life out of your pack without you noticing:

- **Scanning:** This prevents the radio from going into "Power Save" sleep mode. Normally, the radio "pulses" the receiver to check for activity. When scanning, the CPU is running full throttle, checking 3 channels every second. Scanning burns power 3x faster than sitting on one channel.

- **Backlight:** If you have the backlight set to "Always On" (or a long timeout), you are wasting power. Set it to 2 seconds. You don't need a flashlight on your screen just to listen.

- **Dual Watch (TDR):** Checking two frequencies prevents deep sleep. The radio has to wake up twice as often to check Priority A and Priority B.

- **High Power Simplex:** You don't need 5 Watts to talk to your wife in the backyard. Switch to Low Power (1 Watt) by tapping the # key. This cuts your transmit consumption by 60%, significantly extending your "Flamethrower" time.

## SECTION 2: THE MANDATORY UPGRADE (THE USB-C REVOLUTION)

Stop living in 2010. The standard UV-5R charging cradle is a logistical nightmare. It requires 110V AC power (a wall outlet) and a specific "wall wart" transformer. If the power grid goes down, that cradle becomes a paperweight unless you have a generator running. Even if you have a generator, do you really want to use a precious outlet for a single radio charger?

**The Solution: The USB-C Extended Battery.**

Baofeng (and BTech) now sells batteries that have a **direct USB-C charging port** built into the battery casing itself.

**Why This Changes Everything**

1. **One Cable to Rule Them All:** You already charge your phone, your tablet, and your flashlight with USB-C. Now your radio joins that ecosystem. You don't need a special "radio charger" taking up space in your bag. You reduce your cable clutter by 50%.

2. **Charge Anywhere:** You can charge the battery from your car's USB port, your laptop, a solar panel, or, most importantly, a portable power bank. The source doesn't matter, as long as it has a USB port.

3. **Ditch the Cradle:** You can leave the bulky plastic cradle at home. This saves weight and bulk in your Go-Bag. No more worrying about bending the pins on the cradle or losing the AC adapter.

**Action Step:** Go to Amazon or BTech right now. Buy a **BL-5L (3800mAh) USB-C Battery**.

- *Note:* Ensure it specifically says "USB-C Charging" or "Direct Charge." Some older extended batteries look identical (big bricks) but still require the desktop cradle. Read the description carefully.

- *Cost:* ~$20 - $25.

- *Value:* Infinite. This is the single most important hardware upgrade after the antenna. It modernizes a 2012 radio into a 2025 tool.

# SECTION 3: CHARGING STRATEGY (HOME, CAR, BAG)

You need a layered approach to power, just like you have a layered approach to frequencies. Relying on one method is a recipe for silence.

## Layer 1: Home Base (Grid Up)

Keep one radio in the desktop cradle, plugged into the wall. This is your "Grab and Go" unit.

- *The Trap:* Do not leave it in the cradle 24/7/365. Lithium batteries hate being at 100% voltage constantly. It degrades their chemistry, leading to "puffy" batteries or reduced capacity.

- *The Fix:* Use a simple outlet timer (like for Christmas lights) to run the charger for 2 hours a day. Or, simply create a routine where you rotate the radio out of the charger once a week to use it, then put it back.

## Layer 2: The Vehicle (Mobile)

Your car is a massive generator on wheels. It is the ultimate survival battery.

- *The Old Way:* Buying a "Battery Eliminator" (fake battery that plugs into cigarette lighter). These are okay, but they tether you to the car. You can only talk while sitting in the driver's seat.

- *The New Way:* A USB-C cable plugged into your dashboard USB port, charging your USB-C battery. You can charge *while* you drive to the scene, then unhook and walk away with a full tank of electrons.

**Layer 3: The Go-Bag (Grid Down)**

This is where the USB-C strategy shines. Instead of carrying a specialized radio charger, you carry a **standard Anker-style Power Bank** (10,000mAh or 20,000mAh).

**The Math of Survival:**

- Radio Battery = ~1,800mAh (Standard) or ~3,800mAh (Extended).

- Power Bank = 20,000mAh.

- *Loss:* Voltage conversion efficiency is about 80% (some energy is lost as heat).

- *Result:* A single 20,000mAh brick can recharge your standard radio **8 to 10 times.**

- *Endurance:* If one charge lasts 2 days (conservative use), **one brick gives you 20 days of communications.** That is nearly three weeks of autonomy in a package the size of a smartphone.

# SECTION 4: THE "AA" OPTION (THE BREAK-GLASS BACKUP)

Lithium batteries are great, until they are dead. When the Anker brick is empty, the car is out of gas, and the sun isn't shining, what is left? **The AA Battery.**

You can scavenge AA batteries from TV remotes, children's toys, abandoned wall clocks, and junk drawers in almost any house in America. They are the currency of the apocalypse.

**The AA Battery Case (BL-5 AA)**

You can buy a plastic clamshell case (about $8) that snaps onto the back of the UV-5R and takes 6x AA batteries (or 6x AAA depending on the model).

**The Pros:**

- **Scavenge Capability:** You are no longer dependent on the grid. As long as you can find AAs, you can talk.

- **Instant "Recharge":** No waiting for a USB cable. Swap the cells and you are back on air in 30 seconds.

**The Cons (Why this is a BACKUP only):**

1. **Size:** It makes the radio physically huge (about 2 inches longer). It won't fit in standard pouches.

2. **Voltage Sag:** AA Alkaline batteries are not designed for high-current draw. When you press PTT on High Power, the batteries panic and the voltage drops instantly. You might get "Low Voltage" warnings even if the batteries are half full.

3. **Spark Gap:** The fit and finish on these cases is often cheap. If you drop the radio, the case pops open and batteries fly everywhere.

**The Verdict:** Buy one. Put 6 lithium (Energizer Ultimate) AA batteries in it (they have 10-year shelf life and handle cold weather better). Tape it shut with electrical tape to prevent "explosive disassembly" if dropped. Put it in the bottom of your bag. **Do not use it unless you have to.** It is your "Last Stand" power source.


## SECTION 5: SOLAR VS. STORED (DON'T GET FOOLED)

You will see "Solar Chargers" marketed to preppers. They show a guy hiking with a tiny solar panel clipped to his backpack, looking rugged. **This is mostly fantasy.**

### The Solar Reality

Small, backpack-sized solar panels produce very little current, especially if you are moving. Shadows from trees, changing angles, and overcast skies kill the output.

- To charge a 20,000mAh brick with a small "backpack" panel might take **3 to 4 days of perfect, stationary sunlight.**

- Directly charging a radio from a solar panel is often impossible because the fluctuation in voltage (as clouds pass) confuses the radio's charging circuit, causing it to stop charging entirely.

### The Stored Power Reality

A 20,000mAh USB battery brick costs $30, fits in your pocket, and works in a pitch-black basement during a hurricane. It works at night. It works in the rain.

**Tactical Dad Rule:** Focus on **Storage** first, **Generation** second.

1. Buy two high-quality USB Power Banks.

2. Keep them charged.

3. Only mess with solar if you are planning for a "Long Term (30+ day) Grid Down" scenario. For a 3-day or 7-day event, stored power wins every time. It is reliable math vs. variable weather.

# SECTION 6: ENVIRONMENTAL HARDENING (HEAT & COLD)

Batteries are chemical reactions. Temperature controls the speed of the reaction.

### The Cold (The Lithium Killer)

If you leave your radio in your car in Minnesota in January (-10°F), the battery will read "Dead" almost instantly. The chemical reaction slows down so much it can't provide the current needed to transmit.

- *The Fix:* **Body Heat.** Keep the radio inside your jacket, close to your chest. The warmth of your body keeps the chemistry active. Use a remote speaker mic (routed out of your zipper) to talk. Do not leave radios in unheated cars overnight in winter, bring them inside.

### The Heat (The Capacity Killer)

If you leave your radio on the dashboard in Arizona in July (140°F), you are permanently cooking the chemistry.

- *The Result:* The battery won't explode, but it will lose capacity permanently. A battery that used to hold a charge for 24 hours will now only last 12.

- *The Fix:* Store radios under the seat, in a cooler, or in a light-colored bag. Never in direct sun.

# SECTION 7: STORAGE & MAINTENANCE (AVOIDING DRAWER DEATH)

The worst thing you can do to a lithium battery is drain it to 0% and leave it in a drawer for a year. The voltage will naturally drift down over time. If it drops below the "safe" threshold (usually around 2.5V per cell), the battery's internal protection circuit will trip and permanently lock it out. It will never charge again. It is a brick.

### The "Quarterly Rotate" Plan

You don't need to be obsessive. You just need a rhythm. Tie this to another household chore so you remember.

- **When:** The first day of every season (Start of Spring, Summer, Fall, Winter). Same day you check smoke detector batteries.

- **The Drill:**

    a. Turn on every radio. Check voltage (Hold 0 key).

    b. If below 7.4V, charge it up.

    c. Check the Go-Bag power banks. Press the button to check the lights. Top them off if needed.

    d. Inspect the AA alkaline batteries for leaks (white crusty powder). Alkaline leaks can destroy the electronics in your backup case.

# ACTION STEPS & WORKBOOK

## The Power Plan Worksheet

| Kit Location | Primary Power | Secondary Power | Backup Power |
| --- | --- | --- | --- |
| **Home Base** | Desktop Cradle (AC) | USB-C Battery + Wall Plug | Case of AA Batteries |
| **Car (Mobile)** | USB-C Cable (Dash) | Spare BL-5 Battery (Glovebox) | 12V Battery Eliminator |
| **Go-Bag** | USB-C Battery (On Radio) | 20,000mAh Anker Brick | AA Clamshell Case |

## The "72-Hour Comms Readiness" Checklist

*Can you survive a 3-day weekend without power?*

- [ ] **Primary Radio:** Has a USB-C battery installed and fully charged.

- [ ] **Charging Cable:** A USB-A to USB-C cable is IN THE BAG (not borrowed from your phone charger).

- [ ] **Power Bank:** One 10,000mAh+ brick, fully charged (4 lights).

- [ ] **The Backup:** One AA Case loaded with fresh Lithium AA batteries (taped shut to prevent rattling).

- [ ] **Vehicle Check:** Tested the car charger, does it actually work with the engine off? (Some cars kill 12V ports when the key is removed, meaning you can't charge while parked).

# Chapter 8

# "SECURITY" THE HONEST WAY

## PRIVACY, OPSEC, AND WHAT TONES REALLY DO

IF YOU HAVE WATCHED any spy movies, you probably think you can "encrypt" your communications. You imagine pressing a button and having your voice turn into scrambled robot noise that only your team can understand. You picture a specialized headset and a secret frequency that no one else knows exists.

Let me burst that bubble immediately: **On a Baofeng UV-5R, privacy does not exist.**

The airwaves are public property. They are a shared space, like a public park. Anyone with a $25 radio can hear you. Anyone with a $300 police scanner can record you. Anyone with a $20 software-defined radio (SDR) dongle on their laptop can visualize your signal on a map and log every word you say to a hard drive.

If you are transmitting, you are broadcasting.

Security on analog radio does not come from technology. It comes from **behavior**. It comes from what you say, how you say it, and when you choose to stay silent. This is called **OPSEC** (Operational Security). It is the art of denying the enemy information about your intentions and capabilities.

In this chapter, we are going to learn how to hide in plain sight. We are going to stop pretending we are spies and start acting like disciplined operators.

# SECTION 1: THE GREAT TONE MYTH (CTCSS/DCS IS NOT ENCRYPTION)

This is the most dangerous misconception in the hobby. It gets people into trouble because it creates a false sense of security.

You go into the menu. You set a "Privacy Tone" (CTCSS) of 103.5 Hz. You test it with your wife. She can hear you, but when you turn her tone off, she can't hear you (because her squelch is closed). You think: *"Great! We are on a private line! Nobody else can hear us!"*

**The Truth:** CTCSS (Continuous Tone-Coded Squelch System) is not a wall. It is not encryption. It is a filter.

- **What you think it does:** Scrambles your voice so others can't hear.

- **What it actually does:** Tells *your* radio to ignore everyone else. It puts a filter on *your* speaker, not on *their* ears.

**The "Earplug" Analogy:** Imagine you are in a crowded restaurant.

- **Encryption** would be you and your wife speaking a secret language (Elvish). Nobody understands you, even if they listen closely.

- **CTCSS** is you and your wife putting in specialized earplugs that block all sound *except* your specific voices. You can't hear the other diners. You feel like you are in a private bubble. **But the other diners can still hear you screaming across the table.**

**The 'Monitor' Check:** Because your 'earplugs' (Tones) make you deaf to other traffic, you might accidentally talk over someone else who is already using the channel. Always press the **MONI** (Monitor) button on the side of the Baofeng for 2 seconds before you talk. This temporarily pulls out the earplugs (opens squelch) so you can check if the channel is actually clear.

If a bad actor has a scanner, he isn't using tones. He is listening to "Carrier Squelch" (Open). He hears everything, the truckers, the kids playing with walkie-talkies, and you discussing where you hide your spare key.

*Tactical Dad Rule:* Never say anything on a "Privacy Tone" channel that you wouldn't feel comfortable yelling across a crowded Walmart parking lot. If you wouldn't say it to a stranger, don't say it on the radio.

# SECTION 2: THE "SCRAMBLER" TRAP (VOICE INVERSION)

Deep in the Baofeng menu (usually Menu 10 or similar on unlocked models), there is a setting often labeled SCR or SCRAMBLER. When you turn it on, your voice sounds like a duck quacking underwater or a warped vinyl record. It sounds cool. It sounds "tactical."

**The Legal Reality:** Using voice scrambling on Amateur Radio (Ham) frequencies or GMRS frequencies is **illegal**. The FCC rules explicitly state that messages must not be obscured to hide their meaning. The amateur bands are for public, open communication. If you use it, you are breaking federal law.

**The Tactical Reality:** Even if you don't care about the law, it is technically weak security. "Voice Inversion" is 1950s technology. It simply flips the high frequencies to low and low to high.

- **The Vulnerability:** Any determined listener can unscramble it with a simple computer program (like Audacity) or even another Baofeng by guessing the inversion frequency. It is not military-grade AES-256 encryption. It is a screen door, not a bank vault.

**The "SHTF" Exception:** In a true "Without Rule of Law" scenario (zombies, total societal collapse), FCC rules obviously don't apply. In that extreme, life-or-death instance, the scrambler might offer a tiny layer of obscurity against casual listeners who are just scanning for clear speech. But do not rely on it. A sophisticated listener will identify the sound of a scrambler immediately, and it might actually draw *more* attention to you because it sounds like you have something to hide. It is a paper shield.

# SECTION 3: OPSEC (OPERATIONAL SECURITY)

If we can't use technology to hide, we must use discipline. We must assume someone is always listening.

### 1. Channel Discipline (The "Hopping" Plan)

Don't camp on one frequency forever. If you talk on Channel 1 every day at 5:00 PM, you establish a "Pattern of Life." Anyone monitoring will know exactly where to find you.

**The Fix: The PACE Plan.** In the military, we use a PACE plan for communications: **P**rimary, **A**lternate, **C**ontingency, **E**mergency.

- **Primary:** "Channel 1." We use this for routine checks.

- **Alternate:** "Channel 5." We switch here if Channel 1 has interference or too many strangers.

- **Contingency:** "Channel 7." We switch here if we suspect we are being monitored.

- **Emergency:** "Channel 0 (National Call)." We go here if everything else fails.

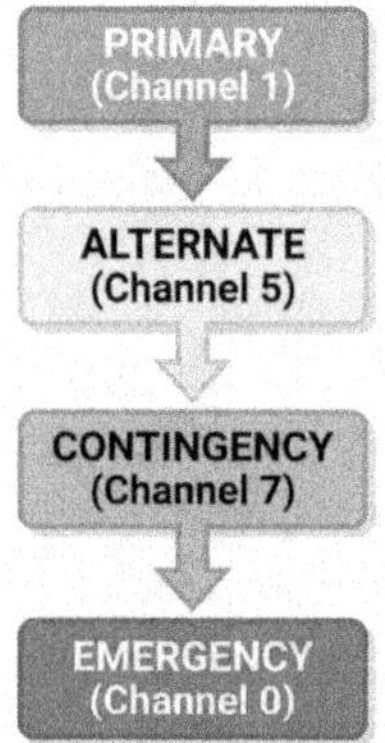


*Implementation:* If you feel uneasy, simply say: "Switch to Blue" (where Blue = Channel 5 in your notebook). Don't say "Switch to 146.580."

## 2. What NOT to Say (The Red List)

Data Aggregation is real. One piece of info isn't dangerous. Ten pieces are deadly. A listener might hear "Mike" on Monday, "Maple Street" on Tuesday, and "Generator" on Wednesday. By Friday, he knows Mike on Maple Street has a generator.

**The Red List (NEVER SAY):**

- **Specific Street Addresses:** "I'm at 123 Maple Drive." Instead say: "I'm at Base."

- **Gate Codes or Lock Combinations:** Never.

- **Detailed Inventory:** "I have 50 gallons of gas and a generator." You just put a target on your house.

- **Dates of Absence:** "We are leaving for vacation tomorrow." You just invited burglars.

- **Full Names of Children:** Use nicknames or tactical callsigns.

## 3. The "Innocuous" Code System

You don't need to sound like a SEAL Team. In fact, sounding like a soldier attracts attention. If you are shouting "Alpha Bravo Tango, perimeter breach!", every listener with a scanner is going to stop and listen to you. You sound interesting. You sound like you have supplies.

You want to sound like a boring suburbanite. You want to be the "Gray Man" of the airwaves.

Instead of using "Tactical Codes" (e.g., "Condition Red"), use **"Innocuous Codes"**, phrases that sound normal but have hidden meaning.

**Family Code Worksheet:**

| The Message | The "Tactical" Way (BAD) | The "Gray Man" Way (GOOD) |
| --- | --- | --- |
| "Come home NOW. Danger." | "Initiate Alpha Protocol! Rally Point!" | "Can you pick up a pepperoni pizza?" |
| "I am being followed." | "Hostile bogie at 6 o'clock!" | "My check engine light just came on." |
| "Switch to backup channel." | "Go to Channel 4! Secure Comms!" | "Call Uncle Mike." |
| "Do not come home. It's unsafe." | "Perimeter breached! Abort!" | "The dog is sick, don't let him out." |
| "I have the supplies." | "Asset secured. In transit." | "I got the milk." |

*Why this works:* A scanner listener hearing "Can you pick up a pizza?" will ignore it. They are looking for excitement. Bore them to death. Hidden in plain sight is better than a lock.

# SECTION 4: LOW-PROFILE COMMS HABITS

The best way to stay secure is to be brief. The longer you talk, the easier you are to find.

## 1. The "Burst" Transmission and Direction Finding

Radio signals can be tracked. This is called "Fox Hunting" or Direction Finding (DF). A person with a directional antenna (Yagi) can swing it around and find exactly where a signal is coming from.

- If you talk for 5 minutes, they can triangulate your exact house.

- If you talk for 5 seconds, they barely have time to grab their radio.

**The Rule:** Keep transmissions under 10 seconds.

- **Bad:** A 5-minute rant about the traffic, the weather, and your feelings.

- **Good:** "Traffic stopped at Exit 5. Detouring. Out." (5 seconds).

## 2. Predictability is Death

Do not set a rigid schedule like "We will talk every hour on the hour." This allows a listener to set an alarm, wake up, and record you. Use a **"Window" system**.

- "I will check in between :15 and :20 past the hour." This forces a listener to monitor you constantly, fighting static and boredom. Most will give up.

## 3. Power Discipline (RF Footprint)

Your radio puts out a signal 'bubble.'

- **High Power (5W):** Big bubble (5-10 miles). Everyone in the county can hear you.

- **Low Power (1W):** Small bubble (1-2 miles). Only your immediate neighbors can hear you.

**The Rule:** Use the lowest power necessary to make the connection. If your team is 1 mile away, switch to Low Power (tap the # key). There is no tactical advantage to broadcasting your conversation to the next county.

# BREAKOUT: THE DIGITAL (DMR) ELEPHANT

You might be asking: *"Why don't we just buy digital radios? I heard they have encryption."*

**The Tech:** Digital Mobile Radio (DMR) converts your voice into 1s and 0s (data). It sounds like modem noise to an analog radio. It can indeed use AES-256 encryption, which is military-grade. Nobody can listen to that without the key.

**Why we aren't doing it in this book:**

1. **Complexity:** Programming a UV-5R is like programming a VCR. Programming a DMR radio (like an Anytone 878) is like programming a Linux server. It requires "Codeplugs," "Color Codes," "Time Slots," "Talk Groups," and "Encryption Keys." The learning curve is a vertical wall. Most people give up.

2. **Incompatibility:** If you have a fancy encrypted radio, you can **only** talk to other people with that exact same radio and encryption key. You cannot talk to your neighbor's Baofeng. You cannot talk to the local ARES volunteers. You cannot shout for help on a standard channel. You isolate yourself in a digital fortress.

3. **Cost:** A good DMR radio is $100-$300. A Baofeng is $25. For a family of four, equipping everyone with DMR is a $1,000 investment versus $100.

4. **Failure Modes:** Digital signals are "all or nothing." Analog signals degrade gracefully, you can still understand a scratchy analog voice. A weak digital signal just disappears.

*Verdict:* Stick to analog for now. Master the basics. Learn to walk before you try to fly a helicopter.

# WORKBOOK TOOLS

**The OPSEC Audit:** *Sit down with your family and fill this out.*

**Our "Green List" (Okay to say):**

- First names only (or callsigns).

- General landmarks ("The grocery store," "The high school").

- Weather reports / Road conditions.

- "I am safe" / "I am moving."

**Our "Red List" (NEVER say):**

- "We have plenty of food/ammo/gas." (Invites looting).

- "The alarm system is broken." (Invites burglary).

- "We are leaving the house empty for three days." (Invites squatting).

- "I am alone."

**The "Pizza Code" Card**

*Create 3 codes that only your family knows. Memorize them. Do not write the meanings on the card, only the phrases.*

1. **The "Get Out" Phrase:** (Means: Leave immediately, meet at rally point. Do not pack, just go).

    ◦ *Phrase:* __________________ (e.g., "Grandma is asking for you.")

2. **The "Duress" Phrase:** (Means: I am being forced to talk, but I am not safe. Someone is with me).

    ◦ *Phrase:* __________________ (e.g., "I'll be home in 100 minutes" - knowing you are only 10 mins away).

3. **The "Silence" Phrase:** (Means: Turn off the radio immediately, stop transmitting, listen only. We are being hunted).

    ◦ *Phrase:* __________________ (e.g., "Goodnight Moon.")

# Chapter 9

# THE FAMILY/TEAM COMMS PLAN

## FROM GEAR TO CAPABILITY

You HAVE SPENT WEEKS learning about antennas, batteries, and repeaters. You are now a competent radio operator. You can hit a repeater 20 miles away and you know the difference between a UV-5R and a UV-82. Congratulations. You are halfway there.

If the grid goes down and you are the only person in your family who knows how to turn on a radio, **you have failed.**

A radio is a team sport. It requires someone to talk *to*. If your wife thinks the radio is just "Dad's weird hobby" and your kids are afraid to touch it because you yelled at them once for pressing the wrong button, you do not have a communications plan. You have a toy collection.

In a crisis, hardware failure is annoying, but **human failure is fatal.** If your family freezes because they don't know the channel, or if they talk over each other because they lack discipline, the expensive radios become paperweights.

This chapter is the hardest part of the book because it involves people, not physics. You can't program people with a cable. We are going to turn your family into a cohesive, capable team.

# SECTION 1: SELLING THE PLAN (WITHOUT SOUNDING CRAZY)

You cannot force your family to care about radio. If you start talking about "EMPs," "Civil War," and "Societal Collapse," their eyes will glaze over. They will label you as paranoid, and they will tune you out. You need to sell **utility**, not fear. You need to show them how this tool makes *their* life easier right now, not just in the apocalypse.

**The "Convenience" Pitch (The Gateway Drug):** Don't say: *"We need this for when the Russians invade."* Say: *"Next time we are at the County Fair or Disney World, you know how the cell service always crashes because there are 50,000 people posting selfies? We can use these to find each other so we don't have to wait in line for food alone. It's basically a VIP line for talking."*

**The "Safety" Pitch (For the Spouse):** *"If a storm knocks out the cell towers like it did last year, I want you to have a way to call me from the house if you get hurt or if the car breaks down. It's just a backup phone. I hope we never use it, but I'll sleep better knowing you have it."*

**The "Cool Factor" Pitch (For the Kids):** *"It's a secret channel. We can talk without Mom hearing us from the backyard."* (Trust me, this works. Kids love exclusivity).

- **For Teenagers (The Texting Rebuttal):** They will say, 'I'll just text you.' You reply: 'Texts get stuck when towers are overloaded. Radios don't. Do you want to be the only one stuck at the concert with no ride home because your text didn't go through?' Appeal to their independence.

# SECTION 2: ROLES & THE "BUS FACTOR" (WHO IS IN CHARGE?)

In most families, Dad is the "Comms Guy." He programs the radios, charges the batteries, and initiates the calls. He is the single point of failure. **The Problem:** If Dad gets hurt, is driving, is stuck at work, or is handling a physical threat, the system collapses. You are the bottleneck.

## The Roles

You need to assign roles based on personality and capability, not just age.

1. **Net Control (The Hub):** Usually you. You direct traffic. You decide which channel the family uses. You prioritize messages. "Unit 2, stand by. Unit 3, go ahead."

   ○ *Responsibility:* Maintains the master battery supply, monitors the repeater, and makes the final call on movement.

2. **The XO (Executive Officer / Second-in-Command):** Usually the spouse or oldest capable child. *This is the critical role.*

   ○ **The Protocol:** The XO monitors the radio when Net Control is busy. If Net Control does not check in within 15 minutes of a scheduled time, the XO assumes command. They make the decision: "Okay, Dad isn't answering. We are switching to Plan B. Everyone meet at the house."

   ○ *Why this matters:* It gives your family permission to act without you.

3. **The Scouts (Rovers):** The kids or mobile units. Their job is simple: Listen and Report.

   ○ *Responsibility:* Keep the radio ON. Do not touch the channel knob. Answer when called. Report what they see ("I see a road block").

## The "Incapacitated Operator" Drill (Morbid but Necessary)

Ask your family this question tonight at dinner: *"If I am unconscious on the floor and the power is out, can you turn on this radio and call for help?"*

If the answer is "I don't know which button is On," or "I don't know what channel the police are on," you have work to do.

• **The Fix:** Put a laminated "Cheat Sheet" on the back of the designated "Kitchen Radio" (the one that stays in the charger).

   a. Turn Top Knob clockwise (ON).

   b. Check Screen says "Channel 1".

   c. Press and Hold the LARGE Side Button (PTT).

d. Speak: "Emergency. Help."

## SECTION 3: YOUR COMMS PLAYBOOK (THE 3 MODES)

You don't wear a raincoat on a sunny day. You shouldn't act like you're in a warzone when you're just parking a trailer. You need different behaviors for different threat levels.

### Mode 1: Normal Day (Green)

- **Status:** Grid is up. Cell phones work. The world is normal.

- **Radio Use:** Recreational only. Hiking, camping, convoy driving to the beach.

- **Goal:** Practice without stress. Use the radios to coordinate backing up the RV or finding a parking spot at the mall.

- **The Benefit:** This builds muscle memory. Your family gets used to holding the button, waiting a second, and speaking clearly. If you only bring the radios out during a disaster, they will fumble.

### Mode 2: Elevated Alert (Yellow)

- **Status:** Severe Weather Watch, Local Power Outage, Civil Unrest nearby, or "The Cell Network is acting weird."

- **Radio Use:** Monitoring and Readiness.

- **Action Steps:**

  - **Net Control:** Turns on the "Home Base" radio (the one with the good antenna) to scan local repeaters, NOAA weather, and emergency frequencies.

  - **The Team:** Ensures all handhelds are placed in their chargers or have fresh batteries inserted.

  - **The Check-In:** "Everyone take a radio. Keep it on Channel 1. Check in once an hour on the hour. Keep it quiet otherwise."

**Mode 3: Emergency Mode (Red)**

- **Status:** Cell towers dead. Disaster declared. Family separated. Immediate threat.

- **Radio Use:** Primary Command & Control.

- **Action Steps:**

  - **Deploy the Big Gun:** Connect the Mag Mount antenna (on a cookie sheet or roof) to the main radio for maximum range.

  - **Initiate "Pizza Code":** (See Chapter 8). Use code words to verify safety.

  - **Net Control:** Stays on the radio 24/7 (or runs shifts with the XO).

  - **Silence Rule:** Unless you have "Traffic" (important info), stay off the air to save batteries and keep the frequency clear. No "Just checking in" unless scheduled.

# SECTION 4: GAMIFIED DRILLS (MAKE IT FUN)

If you look at your kids and say "Time for a readiness drill," they will groan and hide. If you say "Let's play Radio Hide-and-Seek," they will fight over who gets the radio. Gamification bypasses the boredom and plants the skills deep in their brains.

**Game 1: "Radio Hide-and-Seek" (Signal Logic)**

- **The Setup:** Dad is "Base" on the front porch with a radio. Kid is "Rover."

- **The Rules:** The Kid hides somewhere in the neighborhood (within safe boundaries).

- **The Gameplay:** The Kid keys up. *"Base, this is Rover. Can you hear me?"*

  - Dad gives a signal report: *"Loud and clear"* or *"Scratchy, you sound like a robot."*

  - Kid gives a clue: *"I see a red mailbox."*

- **The Physics Lesson:** The kid learns quickly that hiding in a concrete culvert kills the signal, but climbing the playground slide makes it crystal clear. They learn **Line of Sight** intuitively through gameplay, not a lecture.

## Game 2: "The Scavenger Hunt" (Pro-Words & Listening)

- **The Setup:** Mom and Dad stay home. Kids go to the park or woods.

- **The Rules:** Dad gives instructions over the radio one at a time. The kids cannot move to the next target until they confirm via radio.

- **The Gameplay:**

  - Dad: *"Rover, target one is a pinecone. Over."*

  - Kid: *"Wilco. Finding pinecone."* ... (Pause) ... *"Base, I have the pinecone."*

  - Dad: *"Copy. Target two: A smooth rock. Over."*

- **The Lesson:** This teaches **Pro-Words** (Copy, Wilco) and listening skills. If they don't listen effectively, they don't find the treasure. It teaches them to wait for the transmission to end before acting.

## Game 3: "The Zombie Run" (Stress & Silence)

- **The Setup:** A "Grid Down" simulation during a camping trip.

- **The Scenario:** "Zombies" (Dad) are patrolling the woods. Mom and Kids must navigate to a "Safe Zone" (the car) using only the radio to coordinate their movement without getting caught.

- **The Rules:**

  - They must split up (Flanking).

  - They must whisper (Volume discipline).

  - They must use code words ("Zombies are at the fire pit" = Dad is at the campfire).

  - If Dad hears them talking on his radio, he "catches" them (mimics interception).

- **The Lesson:** This teaches **Radio Discipline** (whispering, brevity, codes) and OPSEC under mild adrenaline. It teaches them that the radio is a tool for evasion, not just chatting.

## SECTION 5: AFTER-ACTION REVIEW (AAR)

After every trip, game, or actual storm, ask three questions at dinner. This is not a lecture; it is a debrief. Do not skip this.

1. **What broke?** (Did a battery die unexpectedly? Did a belt clip snap? Did the cheap headset hurt your ears?)

2. **What was confusing?** (Did we mix up Channel 1 and Channel 2? Did you not understand what "Wilco" meant?)

3. **What do we fix?** (Buy a spare battery. Laminated the cheat sheet. Practice the phonetic alphabet more).

*The Golden Rule of the AAR:* **Blame the process, not the person.** If your wife couldn't figure out how to unlock the keypad, don't say "You should have studied." Say "The process failed. We need to disable the keypad lock or make a better checklist."

# WORKBOOK TOOLS

## The Family Comms Card

*Fill this out, laminate it using clear packing tape, and zip-tie it inside every back-pack and go-bag.*

**PRIMARY CHANNEL:** 1 (Home - Simplex) **BACKUP CHANNEL:** 5 (Blue - Alt Simplex) **EMERGENCY (Calling for Help):** Channel 0 (National Call)

**IF LOST:**

1. **STOP.** Stop moving. Sit down.

2. Turn Radio **ON** (Clockwise click).

3. Turn Volume **UP** (Halfway).

4. Hold PTT button, count "1 Mississippi", then say: *"This is [Name]. I am lost. I am at [Landmark]. Over."*

5. Let go and **LISTEN**. Wait 30 seconds before trying again.

**DAD'S PHONE:** 555-0199 **MOM'S PHONE:** 555-0188

## The "Net Control" Handoff Checklist

*For the Spouse/XO. Tape this to the "Base Station" radio.*

- [ ] **Trigger:** Dad is unresponsive for 15 minutes OR Dad is injured.

- [ ] **Take Command:** Pick up the Base Radio.

- [ ] **The Call:** "All stations, this is [Name]. I am now Net Control. Report in."

- [ ] **Roll Call:** Wait for each child/unit to respond.

- [ ] **Write Down:** Write down the location of all units.

- [ ] **Decide:** Do we stay put (Shelter in Place) or move to the Rally Point?

- [ ] **Broadcast:** Give the order clearly. "All units, return to base immedi-ately."

# Chapter 10

# TROUBLESHOOTING LIKE A FIELD TECH

## THE DIAGNOSTIC PLAYBOOK

THE RADIO WORKED YESTERDAY. Today, you press the button, and nothing happens. Or worse, you hear people talking, but they can't hear you.

Most people panic. They assume the radio is "broken." They start mashing random buttons, hoping to stumble upon a fix. They perform a "Factory Reset," wiping out all their carefully programmed channels, only to find the problem is still there. Finally, they throw the radio in the "Junk Drawer of Shame" and assume they aren't smart enough for this hobby.

Stop.

A radio system is a chain. Audio enters the mic > converts to electricity > modulates into RF > travels up the coax > radiates off the antenna > flies through the air.

If the system fails, it is because **one specific link** in that chain is broken. Your job is not to be a wizard; your job is to find the broken link. This is called **Diagnostics**. It is the difference between a user who complains and an operator who fixes.

In this chapter, we are going to turn you into a Field Technician. We will replace guessing with a structured workflow that isolates the problem in seconds.

# SECTION 1: THE 60-SECOND TRIAGE (THE "OODA LOOP")

Before you open a menu, connect a laptop, or start tearing things apart, look at the physical radio. 90% of failures are mechanical, not digital. Run this **OODA Loop** (Observe, Orient, Decide, Act) before you panic.

**The "Tactical Dad" Triage Loop:**

1. **Is it ON?**

   - *The Check:* Turn the volume knob clockwise. Did you feel the physical "click" past the detent? Did the screen light up?

   - *The Trap:* Sometimes the knob gets bumped to "just barely off" inside a backpack. Or, the knob is turned on, but the volume is set so low it *appears* off. Turn it up to 50%.

2. **Is it FED?**

   - *The Check:* Hold the 0 key for two seconds. The screen will display the voltage (e.g., 8.2V).

   - *The Standard:*

     - **8.0V - 8.4V:** Full Charge. Green light.

     - **7.4V - 7.9V:** Functional. Yellow light.

     - **Below 7.4V:** The Danger Zone. The radio might turn on and receive audio, but the moment you press PTT (which demands high current), the voltage will sag, and the processor will reboot or the transmission will cut out. It is effectively dead.

3. **Is it CONNECTED?**

   - *The Antenna Check:* Grab the antenna near the base. Twist it clockwise. Was it loose? A loose antenna can burn out your transmitter because the energy has nowhere to go.

   - *The Mic Check (The K1 Gap):* If using a hand mic, push the plug in hard. The "K1" connector is notorious for wiggling loose by 1mm. This creates a "Zombie State" where the speaker pins connect (you hear static), but the mic audio pin disconnects (you transmit silence). Push it until it snaps. *Field Fix:* If it keeps popping out, use your pocket knife to carefully shave 1mm of plastic off the base of the plug. This allows it to seat deeper into the radio body.

### 4. Is it UNLOCKED?

- *The Check:* Look at the screen. Is there a tiny (Key icon) or L visible?

- *The Symptom:* You press buttons to change the channel, but nothing happens. You panic, thinking the keypad is dead.

- *The Fix:* Press and hold the # (Hash/Key) button for 2-3 seconds until the icon disappears. This feature exists to stop your hip from changing channels, but it often confuses new users.

If those four pass, the hardware is alive. Now we hunt for the software glitch.

## SECTION 2: SYMPTOM FLOWCHARTS (LOGIC TREES)

Don't guess. Follow the path. By asking binary (Yes/No) questions, we eliminate variables.

### Symptom 1: "I Can't Hear Anything" (The Deaf Radio)

*Scenario: You are standing next to your buddy. His radio is chattering away with traffic. Your radio is sitting there in silence.*

- **Step 1: Check the Volume.**

  - Turn the knob all the way up. Sometimes "Low" is actually "Silent" on cheap potentiometers. (Yes > Next).

- **Step 2: Check the Squelch (The Gate).**

  - Press and hold the MONI (Monitor) button on the side (usually the small button below PTT). This bypasses all programming and forces the speaker open.

  - *Result A (Silence):* If you hold the button and hear **absolute silence**, your speaker wire is broken or your audio amplifier is dead. The radio is physically broken. Swap it.

  - *Result B (Static):* If you hear a loud **SSSSHHHHH**, your speaker works. The radio is receiving. The problem is a setting that is keeping the "Gate" closed. > Next.

- **Step 3: Check the Tone Squelch (R-CTCS).**

  - Look closely at the left side of the screen. Do you see a tiny "CT" or "DCS" icon?

  - *The Problem:* You have a "Receive Tone" set. Your radio is politely waiting for a specific password (e.g., Tone 103.5) before it unmutes. The people talking aren't sending that password, so your radio ignores them.

  - *The Fix:* Go to Menu 11 (R-CTCS) and Menu 10 (R-DCS). Turn them both to **OFF**. You want to hear everything.

- **Step 4: Check the Frequency.**

  - Look at the small numbers (e.g., 146.520). Are you on **exactly** the same frequency as your buddy? A difference of 0.005 MHz is enough to make you deaf or make voices sound like "Duck Talk."

- **Step 5: Check the Active Band (The Arrow).**

  - Look for the tiny black arrow (▶) on the left side of the screen. Is it pointing to the Top Row (A) or Bottom Row (B)?

  - *The Trap:* If you are programmed on Top (A), but the arrow is on Bottom (B) and Dual Watch is OFF, you will not hear anything on A.

  - *The Fix:* Press the EXIT/AB button to move the arrow to the correct line.

### Symptom 2: "I Can Hear, But Can't Transmit" (The Mute Radio)

*Scenario: You press PTT to talk. You shout your message. Nobody answers. They continue talking as if you don't exist.*

- **Step 1: Check the Indicator Light.**

  - Press the PTT button. Does the little LED light on the front turn **RED**?

- *NO:* Your PTT button is physically broken or the battery is too dead to fire the transmitter circuit.

- *YES:* The radio is generating RF power. The signal is leaving the building. > Next.

- **Step 2: Check the Offset (Repeater vs. Simplex).**

  - Are you trying to talk "Simplex" (direct) to a friend?

  - Look at the top of the screen. Do you see a tiny + or - icon?

  - *The Problem:* You are listening on 146.520, but when you press the button, the radio jumps to 146.580 (the offset). Your friend is still listening on .520. You are shouting into the void on a different frequency.

  - *The Fix:* Go to Menu 25 (SFT-D) and turn it **OFF**.

- **Step 3: Check the Mic (The Mechanical Failure).**

  - Are you using a headset or hand mic?

  - *The Test:* **Unplug it.** Throw it on the seat. Pick up the radio and talk directly into the built-in mic.

  - *The Result:* If they can hear you now, your headset wire is broken internally. This is the #1 cause of transmit failure in cheap gear. The wires inside those $10 headsets are as thin as hair and break if you look at them wrong.

## Symptom 3: "I Can Hit Simplex, But Not the Repeater"

*Scenario: You can talk to your wife in the driveway perfectly. But when you try to hit the "County Repeater" 10 miles away, you get nothing.*

- **Step 1: The "Kerchunk" Test.**

  - Tap the PTT button for 1 second and release. Listen immediately.

  - *The Sound:* Do you hear a burst of static or a "beep" return immediately after you let go? That is the "Repeater Tail." It means the repeater heard you and kept its transmitter open for a second.

  - *NO (Silence):* You are not opening the repeater. The door is locked.

> Next.

- **Step 2: Check the Encode Tone (T-CTCS).**

  - Almost all repeaters require a "Tone Key" (CTCSS) to open. If you don't send the key, they ignore you.

  - Go to Menu 13 (T-CTCS). Is it set to **OFF**?

  - *The Fix:* Look up the repeater on RepeaterBook. Find the "Tone" (e.g., 103.5). Enter that value into Menu 13.

- **Step 3: Check the Offset Direction.**

  - Are you shifting the right way?

  - *The Problem:* The repeater listens on a frequency *lower* than its output (-), but you are transmitting *higher* (+). You are knocking on the back door instead of the front door.

  - *The Fix:* Check the database. Toggle Menu 25 (SFT-D) between + and -. Try again.

- **Step 4: The 'Reverse' Check.**

  - While on the repeater channel, tap the * SCAN button once (do not hold). An R will appear on the screen.

  - *What it does:* It forces your radio to listen to the *Input* frequency (what you normally talk on).

  - *The Test:* If you hear other people talking clearly while in Reverse mode, but you can't hit the repeater, the repeater might be down (or you are too close to the other person). If you hear nothing, no one is talking.

## SECTION 3: THE "DESENSE" MYSTERY (WHY CITIES KILL RADIOS)

This is a specific, maddening problem that plagues the Baofeng UV-5R because of its low cost. It confuses new users more than anything else.

**The Symptom:** You are standing in the middle of a city. You have a great antenna.

You should be hearing everything. Instead, you hear **nothing**. The radio is totally silent. Occasionally, it might "burp" a blast of static, then go silent again.

**The Cause:** "Front End Overload" (Desensitization). The receiver chip in a Baofeng (RDA1846) is like a very sensitive ear. It listens to everything at once. If you are standing near a powerful Pager Tower, FM Radio Station, or Cell Tower, that strong signal screams into the Baofeng's ear. The radio reacts by "putting its fingers in its ears" (Automatic Gain Control). It turns its sensitivity down to zero to protect itself. **Result:** It can't hear the strong tower (because it's on a different frequency), but now it *also* can't hear your friend whispering on the ham band. It is deafened by the noise floor.

**The Diagnostic Trick: "The Antenna Twist"**

1. Tune to the frequency you want to hear (e.g., NOAA weather or a busy repeater).

2. Unscrew your antenna.

3. Slowly lift the antenna off the connector so the center pin *barely* touches (or disconnects entirely, just hovering near it).

4. **Result:** If you suddenly hear the voice clearly for a split second, your radio was overloaded.

   ○ *Why this works:* By disconnecting the antenna, you acted as an "Attenuator", you blocked the massive noise source (the local tower), allowing the radio to unplug its ears. Because the radio has surprisingly good sensitivity, it can still hear your friend through the tiny gap, but the noise source is too weak to trigger the overload.

**The Fix:** You cannot fix the chip. You have to move away from the interference source (the radio tower on the building next door) or use a better radio (superheterodyne) for city environments.

## SECTION 4: YOUR SPARES KIT (MODULAR SWAPS)

In the field, you don't solder. You don't rewrite code. You **Swap.** If a part fails, you replace the unit and move on. Troubleshooting in the rain is miserable; swapping parts is fast.

**The "Tactical Dad" Field Repair Kit:** *(Keep this in a quart-sized Ziploc bag in your Go-Bag)*

1. **Spare Battery (USB-C):** The #1 failure point is lack of power. A dead battery looks exactly like a broken radio.

2. **Spare Antenna (The "Rubber Duck"):** If you trip and snap your fancy 15-inch whip antenna, or the connector bends, you need a backup. Keep the original stubby antenna in the bag. It's better than nothing.

3. **Spare Mic/Earpiece:** The wires on cheap headsets break internally constantly. Carry a cheap spare. If your main mic dies, you swap it in 10 seconds.

4. **The "Rescue Cable":** A USB programming cable.

    ○ *Why:* Sometimes, if a Baofeng battery dies completely while scanning, the radio's CPU glitches. It might reset to Chinese language or lose all memories. The cable lets you re-upload your "Known Good" image from your laptop to restore the brain.

5. **The "Cheat Sheet":** A printed list of your frequencies. If the radio dies and you have to use a friend's radio, you need to know the numbers to type in.

**The Monthly Maintenance Drill:**

- **Inspect:** Look at the gold contacts on the back of the battery. Are they shiny? If they are green, dull, or blackened, clean them with a **pencil eraser**. The rubber acts as a gentle abrasive to remove oxidation without scratching the gold plating.

- **Tighten:** Check the two tiny screws on the belt clip. Check the SMA antenna connector. Vibration loosens everything. Finger-tighten them.

- **Charge:** Top off the spare battery. Lithium batteries lose a tiny bit of charge every month.

# BREAKOUT: THE PRO-TOOLS (FOR THE NERDS)

You don't *need* these to be effective, but they remove the mystery and let you see the invisible.

## 1. The SWR Meter (Surecom SW-33)

- *What it does:* It measures "Standing Wave Ratio." It tells you if your antenna is actually radiating power into the air, or reflecting it back down the wire into the radio (which creates heat).

- *Why get it:* If you build your own antennas or use mag mounts on a car, this proves they are safe. A high SWR (over 3:1) means your antenna is broken or untuned and will burn out your transmitter.

## 2. The Signal Stick (SDR Dongle)

- *What it does:* A $25 USB stick that plugs into a laptop. It shows a "waterfall" visual of radio signals on the screen.

- *Why get it:* You can visually *see* your radio's signal. You can see if it is "splattering" (transmitting messy noise on adjacent channels), which is a common failure mode for dying Baofengs. It also lets you see if you are transmitting off-frequency.

## 3. The Multimeter

- *What it does:* Checks voltage and continuity.

- *The Hack:* Use the "Continuity" (Beep) mode to test headset cables. Touch one probe to the plug and one to the earpiece wire. Wiggle the wire. If the beeping stops, the wire is broken inside the insulation. This finds "intermittent" failures that drive you crazy.

# ACTION STEPS & WORKBOOK

## Printable Field Flowchart

*Print this small (index card size) and laminate it. Tape it to your radio pouch.*

**START:** Radio won't TX/RX.

1. **Check Power:**

   - Hold 0 key. Is Voltage > 7.4V?

   - **NO:** Swap Battery immediately.

   - **YES:** Go to Step 2.

2. **Check Physical:**

   - Is Antenna tight? Is Mic plug fully seated?

   - **NO:** Tighten/Push.

   - **YES:** Go to Step 3.

3. **Check Audio:**

   - Press MONI button. Do you hear static?

   - **NO:** Audio Amp dead. Swap Radio.

   - **YES:** Go to Step 4.

4. **Check Settings (The "Deaf" Fix):**

   - Are Tone Squelch icons (CT/DCS) visible on screen?

   - **YES:** Turn Menu 10 & 11 to OFF.

   - **NO:** Go to Step 5.

5. **Check Transmission (The "Mute" Fix):**

- Press PTT. Does RED light turn on?

- **NO:** Check PTT button / Battery.

- **YES:** Check Offset.

    - Simplex? Ensure + / - icons are GONE (Menu 25 OFF).

    - Repeater? Ensure + / - matches RepeaterBook.

6. **Still Dead?**

- Perform **Antenna Twist Test** (Check for City Overload).

- Unplug Hand Mic (Check for bad wire).

- **Last Resort:** Menu 40 (RESET ALL). *Warning: Deletes everything.*

## Monthly Maintenance Checklist

- **Physical Check:** Antenna connector straight? Belt clip screws tight? Volume knob smooth?

- **Battery Check:** Voltage check on primary radio + spare battery in bag.

- **Contact Cleaning:** Use pencil eraser on battery contacts.

- **Audio Check:** Call a friend/family member. Ask: "How is my audio? Am I scratchy? Am I quiet?"

- **Program Check:** Did you accidentally delete a channel? Scroll through your zones to verify memory map.

## Chapter 11

# THE UPGRADE PATH

## WHEN TO KEEP THE BAOFENG, WHEN TO STEP UP

YOU HAVE MASTERED THE Baofeng. You have a working channel plan, you know how to troubleshoot a dead mic, and you have made contact with a repeater 20 miles away. You feel competent. You feel like an operator.

But you have also noticed the cracks.

You noticed that when you drive near the hospital or a cell tower, your radio goes deaf right when you need it most. You noticed that the scanning speed is painfully slow, you miss entire conversations while the radio chugs through your list of 20 channels. You dropped it on the driveway, and the battery latch cracked, requiring a tactical application of duct tape to keep the power on.

You are ready for the next level.

The temptation is to go online and buy the "Baofeng Pro Max Ultra V2" with 10 Watts, a camouflaged case, and a gold-plated keypad. **Don't do it.** Buying a slightly more expensive Baofeng is like putting a spoiler and racing stripes on a golf cart. It might look cool, but underneath, it is still a golf cart. It has the same engine, the same flaws, and the same limitations.

This chapter is your roadmap out of the "Budget Tier." We are going to look at upgrades that actually increase your **capability**, not just your credit card bill. We will focus on tools that solve specific problems.

# SECTION 1: THE CHEAP UPGRADES (MAXIMIZING THE $25 RADIO)

Before you drop $200 on a new radio, make sure you have wrung every ounce of performance out of the one you have. Often, a $20 accessory fixes the problem you thought required a new radio.

### The $20 Upgrade: The Signal Stick Antenna

- *The Gain:* Better range, less durability anxiety.

- *The Why:* As discussed in Chapter 5, the stock antenna is an electrical compromise. It is often untuned, meaning a portion of your battery power is converted into heat inside the antenna coil rather than radio waves in the air.

- *The Pick:* The **Super-Elastic Signal Stick**. It is made of Nitinol (memory alloy). You can tie it in a knot, stuff it in a cargo pocket, or bend it in half in your go-bag. When you pull it out, it snaps straight instantly. Unlike the stock antenna, it won't kink. Unlike the heavy "tactical folding" tape-measure antennas, it is lightweight and won't snap the SMA connector off your radio circuit board.

### The $25 Upgrade: The Extended Battery (USB-C)

- *The Gain:* Endurance and charging flexibility.

- *The Why:* The stock charging cradle is a logistical choke point. It requires 110V AC wall power. If the grid is down, or if you are in your truck, that cradle is a paperweight unless you have an inverter.

- *The Fix:* A 3800mAh extended battery with a **direct USB-C port**. This changes the radio from a "toy" to a "tool." You can charge it in your truck, from a solar panel, or from the same Anker power bank you use for your phone. It effectively gives you infinite field endurance without carrying a dedicated charger brick.

### The $15 Upgrade: A Better Mic

- *The Gain:* Audio clarity and command presence.

- *The Problem:* The cheap "earbud" style earpieces included with the radio break internally if you look at them wrong. The stock internal mic picks up wind noise easily.

- *The Pick:* The **CommMountain** or **BTech** heavy-duty speaker mic. It has a larger speaker diaphragm (louder, richer audio), a clickier PTT button (tactile feedback even with gloves), and a thicker, coiled cable that resists fraying. Clipping this to your collar keeps the radio protected on your belt while keeping the audio near your ear, essential for "Gray Man" operations where you don't want to be holding a radio in your hand.

## SECTION 2: WHEN YOU HAVE OUTGROWN THE BAOFENG

How do you know it's time to spend real money? When the **hardware** limits your **mission**. If you find yourself saying, "I can't hear them because of the interference," or "I'm afraid to take this in the rain," it is time to upgrade.

### The "Superheterodyne" Factor (The Big Reason)

This is the technical term you need to know. It is the single biggest difference between a $25 radio and a $150 radio.

- **The Baofeng (Direct Conversion Receiver):** This uses a "System on a Chip" (RDA1846). Imagine you are trying to listen to a friend whisper in a crowded stadium. A Direct Conversion chip hears *everything*, the whisper, the shouting match

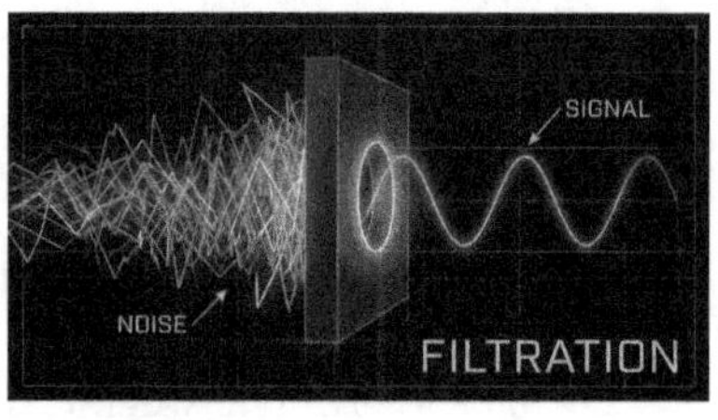


next door, the band playing, and the traffic outside, all at once. It tries to filter it digitally, but if the overall noise is too loud, the chip gets overwhelmed ("Desense"). It effectively goes deaf to protect itself.

- **The Upgrade (Superheterodyne Receiver):** This is like wearing industrial noise-canceling headphones that physically block out every sound frequency *except* the exact pitch of your friend's whisper. It uses physical filters (crystals and coils) to reject interference *before* it hits the brain of the radio.

**You need a Superheterodyne radio if:**

1. **You live in a dense urban environment (RF Soup):** If you are near cell towers, hospitals, or paging transmitters, a Baofeng is useless. It will be silent even when people are talking. A Superhet will hear clearly.

2. **You work Public Service events:** If you volunteer for marathons, bike races, or parades where 50 people are transmitting nearby on different frequencies, a Baofeng will lock up.

3. **Life-Safety Reliability:** If you are relying on this radio to call for a medevac or fire support, you don't want it to go deaf just because you parked next to a radio tower.

### Durability and Water

The UV-5R is "splash resistant" at best. If you drop it in a creek, it dies. If you drop it on concrete, the battery latch often shears off, ejecting the battery and resetting the radio. **You need a better radio if:**

1. **Water:** You kayak, fish, or hunt in the rain.

2. **Impact:** You are clumsy (or your kids are).

3. **Environment:** You want a radio that can survive being tossed into the bed of a truck with firewood.

# SECTION 3: THE GOLD STANDARD HANDHELDS (BUY ONCE, CRY ONCE)

If you are going to spend $150+, buy a radio that will last 20 years. Do not buy midway "upgrades" that are just rebranded Baofengs with a different plastic shell.

### The Tank: Yaesu FT-60R (~$150)

This is the AK-47 of ham radio. It was released in 2004 and is still sold today because it is perfect.

- **Pros:** True Superheterodyne receiver (bulletproof in cities). Incredibly durable (people run them over with trucks and they still work). It includes an AA battery shell, so you can power it with Energizers bought from a gas station in an emergency. Huge aftermarket support.

- **Cons:** Heavy. Old-school NiMH battery (not Lithium), so it discharges faster on the shelf if left for months. No USB charging out of the box (requires a specific cable). The screen is small and orange.

- **Verdict:** The ultimate survival radio. It just works. If the world ends, this is the radio you want in your hand.

### The Modern Workhorse: Wouxun KG-UV9D Mate (~$180)

- **Pros:** Superheterodyne. Large color screen. Holds a huge battery. **Cross-Band Repeat:** This is a killer feature. You can leave this radio in your car (connected to a mag mount antenna on the roof) and use a small Baofeng to talk to it. It will receive your weak signal and "repeat" it out to the world with 10 watts of power. It turns your car into a mobile repeater station.

- **Cons:** The menu system is complex. It has a learning curve compared to the simplicity of the Yaesu.

- **Verdict:** The best "modern" analog handheld for power users who want features, performance, and the ability to extend their range via cross-band repeat.

### The GMRS Specialist: Wouxun KG-935G (~$170)

- **Pros:** IP66 waterproof (can take a hose spray or heavy rain). Specifically tuned for GMRS frequencies. Incredibly loud speaker (cuts through road noise or wind). Simple, locked-down interface that is hard to mess up.

- **Cons:** Locked to GMRS frequencies (legal requirement). You cannot transmit on Ham frequencies.

- **Verdict:** If your family uses GMRS exclusively (no Ham licenses), this is the Rolls Royce. It feels like a Motorola police radio in your hand.

## SECTION 4: THE "MOBILE" LEAP (THE SMARTER SECOND PURCHASE)

Most people buy a Baofeng, then immediately buy a better handheld. This is often a strategic mistake. **Your second radio should be a Mobile Radio (Truck Radio).**

A handheld is a tactical compromise (low power, poor antenna location). A mobile radio is a strategic asset.

## Why a Mobile?

- **Power:** Handheld = 5 Watts. Mobile = 50 Watts. You are **10x louder**. You can hit repeaters 50-70 miles away that your handheld doesn't even know exist. You can punch through trees and terrain that stop a Baofeng cold.

- **Antenna:** It uses a dedicated antenna on the roof, solving the "Faraday Cage" issue permanently. Height + Power = Range.

- **Power Source:** It runs off your car battery. You have weeks of runtime (if the engine runs) compared to hours on a battery pack.

## The Recommendations

- **The Budget Beast: Midland MXT400 (GMRS) / Retevis RT95 (Ham).** Simple, 25-50 watts, no screen clutter. Rugged and cheap (~$150). Great for mounting under a seat or in a glovebox.

- **The Gold Standard: Icom IC-2730A (Ham).** A dual-band workhorse with a **detachable faceplate**. You can mount the heavy radio body under the seat and put the small screen/knobs on your dashboard using a suction cup or vent mount. It scans insanely fast and allows you to monitor two frequencies simultaneously (Dual Receive). (~$300).

**The Tactical Dad Strategy:** Put the 50-Watt mobile in the "Family Escape Vehicle" (the SUV or Truck). If you have to evacuate, that radio is your lifeline to the outside world when cell towers fail. Keep the Baofengs for "last mile" scouting when you have to leave the vehicle.

# SECTION 5: HOW TO UPGRADE WITHOUT WASTING MONEY

Don't end up with a drawer full of incompatible junk. The "hidden cost" of upgrading is often the accessories.

**1. Connector Standardization (The Gender Trap)** Radio manufacturers do not agree on connectors.

- **Baofeng:** Uses **Kenwood K1** audio plugs (2-pin) and **SMA-Female** antennas (Radio has a pin).

- **Yaesu/Icom:** Use **Single-pin** or specialized audio plugs and **SMA-Male** antennas (Radio has a hole).

- **The Trap:** Your collection of Baofeng mics and antennas *will not fit* a Yaesu. You cannot just swap them.

- **The Fix:** Factor the cost of a new mic and antenna into your purchase. Or, buy **SMA-Male to SMA-Female adapters** (use sparingly, as they weaken the signal). When you pick a "System" (e.g., Yaesu), stick with it so your accessories transmit across your fleet.

**2. The "Buy Once" Rule** It is better to have **one** $150 radio that works 100% of the time than **six** $25 radios that work 50% of the time. Start with one quality upgrade for yourself (Net Control). Keep the Baofengs as loaners/backups for the kids, neighbors, or cache supplies. Do not buy a "slightly better" Chinese radio hoping it will be "good enough." Make the jump to the Big Three (Yaesu, Icom, Kenwood) or high-end Wouxun and never look back.

## SECTION 6: THE LONG GAME (COMMUNITY & LICENSING)

Gear is fun, but skills save lives. A 50-watt radio is useless if you don't know who to talk to.

### Get Your General License

The Technician license (Level 1) gets you local VHF/UHF. This is "Line of Sight." The **General license** (Level 2) gets you HF (High Frequency). This allows you to talk **around the world** by bouncing signals off the ionosphere.

- *Why:* In a regional disaster (hurricane wiping out the whole state), VHF can't reach help because the local repeaters are down or out of power. HF can reach a guy in Ohio or Texas who can call the Coast Guard for you. It is the ultimate grid-down backup.

**Join a Local Net (Without Being Weird)**

Find a local "repeater net" (a weekly on-air meeting). Look on RepeaterBook or local club websites for schedules.

- **Don't:** Barge in asking about "SHTF scenarios," "Zombie defense," or "what gun is best." You will be ignored and labeled as "that guy."

- **Do:** Listen first. Check in when asked. Say "This is [Callsign], checking in, no traffic."

- **The Value:** These guys know where the dead spots are in your county. They know which repeaters have battery backup and which ones die in an hour. They are your local intelligence network. "Elmers" (mentors) love helping new guys who are humble, polite, and willing to learn.

# WORKBOOK TOOLS

**The Upgrade Decision Matrix**

*Check the boxes that apply to you to find your next purchase.*

- [ ] **The Urbanite:** I live in a dense city center or near a radio tower. My Baofeng often goes silent. (**Need: Superheterodyne Handheld like Yaesu FT-60R**).

- [ ] **The Commuter:** I need to communicate from inside a vehicle to a base > 5 miles away. I spend a lot of time in my truck. (**Need: 50W Mobile Radio**).

- [ ] **The Outdoorsman:** My radio will definitely get wet, dropped in mud, or abused. (**Need: IP67 Waterproof Radio like Wouxun KG-935G**).

- [ ] **The Park Dad:** I just want to talk to my kids in the park or at the mall. (**Stick with Baofeng. Spend money on better antennas**).

## The "Next 30 Days" Practice Plan

- **Week 1 (Programming):** Program your radio with the "Chapter 2" Channel Plan using CHIRP. Ensure you have a "Simplex" home channel and 3 local repeaters programmed with correct offsets and tones.

- **Week 2 (Range Lab):** Conduct the "Range Reality Test" (Chapter 5). Map your coverage from your house to the grocery store, the school, and the highway on-ramp.

- **Week 3 (Family Drill):** Teach your family the "Pizza Code" (Chapter 8) and run the "Hide and Seek" drill (Chapter 9). Get them comfortable pressing the button and speaking clearly.

- **Week 4 (The Live Test):** Check into a local net. Listen for the start time, wait for the "Visitor Check-In," and put your callsign out there. Break the ice.

# CONCLUSION

## YOU ARE NOT A BAOFENG OWNER; YOU ARE AN OPERATOR

You have reached the end of the manual. By now, your radio should be programmed, your batteries charged, and your family briefed on the "Pizza Code." You have the hardware. You have the frequency lists. You have the cheat sheets.

But owning a hammer doesn't make you a carpenter. And owning a radio doesn't make you a communicator.

The difference isn't the gear. It's the mindset. It is the shift from "having" a thing to "being" a capability. This conclusion is not just a summary; it is a final calibration of your expectations.

### THE TALE OF TWO DADS

Let me tell you a story about two men who live on the same street in the suburbs of Atlanta. Both men consider themselves "prepared." Both men bought the exact same Baofeng UV-5R off Amazon three years ago.

**Dad A: The Gear Collector** Dad A feels good about his purchase. He put the radio in his "SHTF Box" in the back of the closet, right next to the freeze-dried food and the survival knife. He left it in the original packaging because he wanted it to be "fresh" for the apocalypse. He downloaded a frequency list from the internet but never actually programmed it because the software looked hard and he promised himself he'd "get to it next weekend." He assumes that when the power goes out, he will turn it on, and it will just work like a cell phone.

Fast forward three years. A severe ice storm hits North Georgia. Power lines snap. The cell towers lose battery backup after four hours. The house is dark, the temperature is dropping, and his wife is stuck somewhere on the highway. Dad A tears open the box by candlelight. He tries to turn the radio on. Nothing. The battery is dead from three years of self-discharge. He frantically digs for the AA battery pack and spills the alkaline batteries on the floor, cursing as he realizes one has leaked white crust inside the case, but manages to scrape it clean and get it to light up. He hears static. He presses the PTT button to call his wife. Nothing happens. He didn't set the offset. He didn't check the frequency.

He shouts into a dead microphone while his kids look at him, scared and cold. His wife asks, "Does it work?" and he has to admit, "I don't know." He owns a radio, but he has zero capability.

**Dad B: The Operator** Dad B isn't a Navy SEAL. He's an accountant. But every Sunday night while taking out the trash, he turns his radio on for 10 minutes. He programmed his local repeater and his simplex home channel using CHIRP. He taught his 10-year-old son how to do a radio check and taped a laminated "Cheat Sheet" to the back of the battery.

When the same ice storm hits, Dad B doesn't panic. He knows the cell towers are failing because he heard the chatter on the local ARES repeater 20 minutes ago. He initiates 'Yellow Alert' status and calls his wife on the "Home" channel before the phones die (she knew to turn her radio on because of the protocol). "Traffic is gridlocked on Hwy 9. Stay put at the store. I am monitoring Channel 1." His wife doesn't panic because she has practiced this. She stays warm in the store. Dad B monitors the weather alerts on his radio and knows exactly when the road crew clears the main street. He guides her home safely using side roads he confirmed were open via radio reports.

Dad B didn't have better gear. He had **proficiency**.

You have spent this entire book becoming Dad B. Do not let that skill atrophy.

## THE WEEKLY RHYTHM (15 MINUTES TO STAY SHARP)

Radio skills are perishable. If you don't use it, you lose it. Batteries die. Frequencies change. Settings get bumped in your bag. The "Set it and forget it" mentality is a trap that leads to failure.

You need a rigid schedule. If you wait for "when I have time," it will never happen.

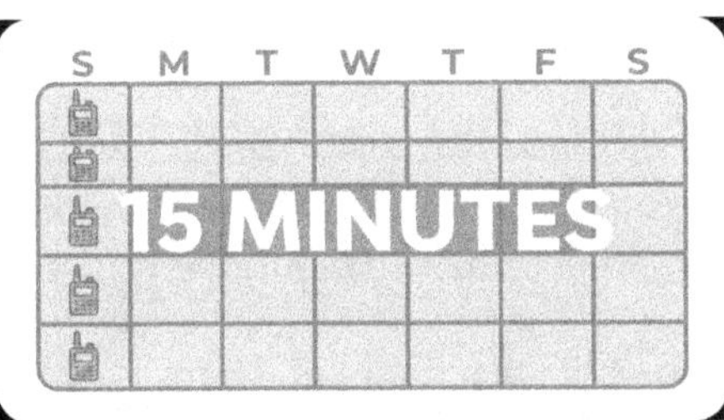


**The "Sunday Night Protocol" (1900 Hours)** *Tie this to a habit you already have, like taking out the trash or prepping coffee for Monday.*

1. **Power Up (2 Minutes):** Turn on every radio in your fleet. Check the voltage (Hold 0 key).

   - *The Standard:* If it's below 7.6V, put it on the charger immediately. If it's fully charged (8.2V+), rotate it off the charger. Lithium batteries hate sitting at 100% on a charger for months; it kills their capacity.

2. **The Physical Audit (3 Minutes):**

   - Check the belt clip screws (they vibrate loose over time).

   - Check the antenna connector (is it tight?).

   - Check the volume knob (is it smooth, or did it get bent?).

   - *Why:* Finding a broken belt clip now is annoying. Finding it while running to a shelter is a disaster.

3. **The "Live" Check (5 Minutes):**

   - **Receive Test:** Tune to the local NOAA weather station. Verify you can hear clear audio. This proves your speaker works and your antenna is receiving.

   - **Transmit Test:** Tune to your primary Repeater. Press PTT for 1 second ("Kerchunk"). Listen for the tail (the return static/beep). This proves your transmit circuit, tone, and offset are still correct.

4. **The Family Drill (5 Minutes):**

   - Hand a radio to your kid/spouse in the other room.

   - **You:** "Base to Unit 1. Radio Check."

- ○ **Them:** "Unit 1. Loud and Clear."

  - ○ *Note:* This isn't just testing the radio; it's testing **them**. It keeps them comfortable with the PTT button so they aren't afraid of it in an emergency. It normalizes the sound of your voice over RF.

5. **Stow (3 Minutes):** Turn them off. Return them to the "Ready Rack" or Go-Bag. Verify the spare battery is present and the "Cheat Sheet" is still attached.

**Total Time:** 15 Minutes. **Result:** 100% confidence that your lifeline works.

# THE FINAL EXAM: EARN YOUR "OPERATOR" STATUS

You have read the book. Now prove you can do the work. Complete these three scenarios. Once you check these boxes, you are no longer a "newbie." You are an asset to your family.

### Scenario 1: The "Blackout" Simulation (Simplex Mastery)

**The Setup:** Turn off the main breaker to your house (or just pretend). Unplug your Wi-Fi router. Put your cell phones in a drawer on "Airplane Mode." **The Mission:**

1. Give a radio to a family member and send them to the end of the block (or a nearby park).

2. Establish contact on your **Simplex Home Channel** (e.g., Channel 1).

3. Have them read a random sentence from a book.

4. You must write it down perfectly. If you miss a word, ask for a repeat using pro-words ("Say Again").

5. **The Twist:** Switch to Low Power (1 Watt) by tapping the # key. Can you still hear them? If not, move to the second floor or stand on a chair. **The Lesson:** You will learn that "Height is Might." You will see exactly how walls and terrain eat your signal, and how moving five feet can change "static" to "clear." You learn that 1 Watt from the second floor beats 5 Watts from the basement.

**Scenario 2: The "Listening Post" (Scanner Mastery)**

**The Setup:** A Saturday afternoon. **The Mission:**

1. Go to RepeaterBook and find the frequency for your local Fire Dispatch or EMS (if analog) or a busy Ham repeater.

2. Program it into your radio using the keypad (Field Rescue Method), no computer allowed.

3. Listen for 30 minutes.

4. **The Goal:** Write down three specific pieces of information you heard (e.g., "Fire truck dispatched to 123 Main St," "Operator KB4XYZ is testing a new antenna," or "Accident at Mile Marker 42").

5. **The Lesson:** This proves you can extract **intelligence** from the airwaves. In a crisis, information is as valuable as water. Knowing where the fire is allows you to drive the other way.

**Scenario 3: The "Grid Down" Message (Repeater Mastery)**

**The Setup:** Drive 5 miles away from your home. Park your car. **The Mission:**

1. Access your local repeater.

2. Wait for a break in traffic. Listen first!

3. Key up and say: "This is 'Your Callsign,' testing a new mobile setup. Can I get a signal report?"

4. **The Goal:** Get a stranger to respond with "Loud and clear."

5. **The Lesson:** This proves your signal can leave your immediate neighborhood. You have touched the outside world. You are not alone. It also breaks the psychological barrier of "Mic Fright." Speaking to a stranger on the air is the final hurdle to becoming an Operator.

## FINAL CALL TO ACTION

You bought this radio because you felt a responsibility. You looked at the fragile state of the world, the storms, the outages, the uncertainty, and you decided to do something about it. You decided to be the one who stays connected when the lights go out.

That is a noble instinct. Protect it.

Don't let this radio become just another piece of plastic in a drawer collecting dust. Keep it charged. Keep it programmed. Keep your family practiced.

When the phones die, and the internet goes dark, and the rest of the world is deaf and blind... **you will be the one with the answers.**

**Base, this is Tactical Dad. The net is open. Out.**

Jared Johnson (KF0RTU)

Jared@MorseCodePublishing.com

# BUILD YOUR LIBRARY

# THE OPERATOR'S UPGRADE PATH

You have mastered the Baofeng. You have the hardware, the channel plan, and the mindset. To truly secure your family's connection to the outside world, you need to expand your legal privileges, your range, and your hardware options.

Here are three guides from **Morse Code Publishing** designed to take you from "Guy with a Radio" to "Communications Commander."

## 1. GET LEGAL, GET LOUD

### Ham Radio Technician Class Study Guide

You know how to use the radio now. But do you have the credentials to use it openly?

Operating "under the radar" is fine for the apocalypse, but it prevents you from practicing on Tuesday afternoons. To truly test your range, join local repeater nets, and build a network of skilled friends, you need your Technician License.

Don't let the exam scare you. You don't need a degree in electrical engineering. This guide strips away the confusing jargon and focuses on the no-nonsense facts you need to pass.

**The Hook:** Stop worrying about the FCC and start transmitting with authority. **Best For:** The operator who wants to join the local ARES/RACES emergency groups.

## 2. WHEN LOCAL ISN'T ENOUGH

## Ham Radio General Class License Study Guide

Your Baofeng is a "Line of Sight" tool. It dominates the neighborhood. But what happens if a hurricane wipes out the power grid for the entire state? What if the local repeaters are dead?

You need to reach the outside world.

Upgrading to your **General Class License** unlocks the High Frequency (HF) bands. This is the "Big League" of radio—bouncing signals off the ionosphere to talk to Texas, Maine, or Europe without satellites or the internet. This guide takes the intimidation out of the General exam, breaking down the math and theory into plain English.

**The Hook:** Break the line-of-sight barrier. Secure your link to the outside world.

**Best For:** The prepared citizen who needs regional and national reach.

## 3. THE "NO-TEST" FAMILY SOLUTION

## GMRS Radios: The Visual Command System

Let's bc honest: You might be willing to study for a radio exam, but your spouse and teenagers probably aren't.

**GMRS (General Mobile Radio Service)** is the answer. One $35 license covers your **entire immediate family**—no test required.

This manual is the perfect companion to your Baofeng setup. It focuses on high-power mobile units for your vehicles and simplified protocols for the family. It teaches you how to build a 50-watt mobile "base station" in your truck and how to integrate GMRS into your family's safety plan without the friction of exams.

**The Hook:** High-power comms for the whole tribe, zcro studying required.

**Best For:** The family fleet, road trips, and overlanding.

# APPENDIX A

## THE ONE-PAGE CHEATSHEETS

ACCESS PRINTABLE VERSIONS OF all appendix documents here:

## CHEATSHEET 1: THE "NO-PANIC" RESET

### IF THE RADIO ACTS WEIRD:

1. **Check Battery:** Is voltage > 7.4V? (Hold 0).

2. **Check Antenna:** Is it tight?

3. **Check Tones:** Are "CT" or "DCS" icons on the left? Turn them OFF.

4. **Check Offset:** Is there a + or - at the top when you are trying to talk Simplex? Turn Menu 25 OFF.

### THE NUCLEAR OPTION (FACTORY RESET):

- *Warning: Erases all channels.*

- Menu 40 (RESET) -> ALL -> Press MENU.

- Change Language back to English: Menu 14 -> ENG.

# CHEATSHEET 2: THE TWO BRAINS (VFO vs. MR)

## THE GOLDEN RULE:

- **VFO (Frequency Mode):** Used for **Programming** and **Searching**. (Manual input).

- **MR (Channel Mode):** Used for **Talking** and **Scanning**. (Saved presets).

## HOW TO SWITCH:

- Press the orange VFO/MR button on the front.

- **IF LOCKED:** Turn radio OFF. Hold MENU button. Turn ON. (Some models).

## VISUAL CHECK:

- **CHANNEL MODE:** You see small numbers (001, 022) on the right side of the screen.

- **FREQUENCY MODE:** No small numbers on the right. Just the big frequency.

## TROUBLESHOOTING:

- *"I can't type in a new number!"* > You are in **Channel Mode**. Switch to VFO.

- *"I can't change the channel!"* > You are in **VFO Mode**. Switch to MR.

- *"Keypad won't work!"* > Check for Lock icon. Hold # key for 2 seconds.

## CHEATSHEET 3: REPEATER MATH (OFFSETS & TONES)

**THE LOGIC:** Repeaters listen on one frequency (Input) and talk on another (Output). You must tell your radio how to knock on the door.

**THE 4 VARIABLES:**

1. **RX Freq:** What you listen to (The big number on screen).

2. **Offset:** How far to shift (Distance).

3. **Direction:** Which way to shift (+ or -).

4. **Tone:** The password key (CTCSS/DCS).

**MENU SHORTCUTS (UV-5R):**

- **Menu 13 (T-CTCS): TRANSMIT TONE.** (The Key). Set this to the repeater's PL Tone (e.g., 103.5).

- **Menu 11 (R-CTCS): RECEIVE TONE.** (The Filter). **KEEP THIS OFF** unless you have a specific reason. It makes you deaf to traffic that doesn't have the tone.

- **Menu 25 (SFT-D): SHIFT DIRECTION.** (+ / - / OFF).

- **Menu 26 (OFFSET): SHIFT AMOUNT.** (0.600 for 2m / 5.000 for 70cm/GMRS).

**THE "REVERSE" TRICK:**

- Press the * SCAN button quickly (tap).

- If you see an R on screen, you are listening to the **INPUT**.

- *Use:* If you hear the person clearly in Reverse mode, you are close enough to talk Simplex.

## CHEATSHEET 4: SCANNING & SQUELCH

### SCANNING RULES:

- **Start Scan:** Hold * SCAN button for 2 seconds.

- **Stop Scan:** Press EXIT or PTT.

- **Skip Channel:** In CHIRP, mark "Skip". On radio, delete channel.

### SCAN MODES (MENU 18):

- **TO (Time):** Pauses for 5 seconds, then moves on (even if they are talking). *Bad for convos.*

- **CO (Carrier):** Pauses while signal is present. Resumes when signal drops. *Best for monitoring.*

- **SE (Search):** Stops when it finds a signal and stays there. *Good for finding new stuff.*

### SQUELCH (MENU 0):

- **0:** Open (Static).

- **1:** Sensitive (Weak signals). *Standard Setting.*

- **5-9:** Insensitive (Strong signals only). *Use in city/interference.*

### FM RADIO:

- Press side CALL button once. (Listen to music).

- *Note:* If a signal comes in on your main channel, the radio effectively pauses the music so you can answer.

# APPENDIX B

## REPEATER WORKSHEET & TRAVEL LIST

**INSTRUCTIONS:**

1. Go to RepeaterBook.com or use the app.

2. Fill this data out **in pencil** (repeaters change).

3. Keep this folded inside your radio case or taped to your sun visor.

4. **Crucial:** The "Input Tone" is the key to the door. Without it, the frequency is useless.

### PART 1: THE "HOME BASE" LIST (CRITICAL INFRASTRUCTURE)

*These are the repeaters covering your daily life: Home, Work, School, and the commute in between.*

| | |
|---|---|
| _______________________ | _______________________ |
| _______________________ | _______________________ |
| _______________________ | _______________________ |
| _______________________ | _______________________ |
| _______________________ | _______________________ |
| _______________________ | _______________________ |

## PART 2: THE TRAVEL / CONVOY LIST

*Fill this out before a road trip. Look for repeaters along your interstate route every 30-50 miles.*

TRIP DESTINATION: _______________________ DATE: __________

TRIP DESTINATION: _______________________ DATE: __________

TRIP DESTINATION: _______________________ DATE: __________

TRIP DESTINATION: _______________________ DATE: __________


## PART 3: THE SIMPLEX STANDARD (NO REPEATER NEEDED)

*Your pre-agreed frequencies for direct radio-to-radio talk.*

___________________________     ___________________________

___________________________     ___________________________

___________________________     ___________________________

___________________________     ___________________________


## QUICK REFERENCE: FREQUENCY BANDS

- **2 Meter Ham:** 144.000 – 148.000 MHz

- **70cm Ham:** 420.000 – 450.000 MHz

- **GMRS:** 462.550 – 462.725 MHz (Channels 15-22)

- **MURS:** 151.820 – 154.600 MHz (Green Dot / Blue Dot)

- **NOAA Weather:** 162.400 – 162.550 MHz (Receive Only)

# APPENDIX C

## CHIRP WORKFLOW CARD & BACKUP LOG

**INSTRUCTIONS:** Use the "Backup Log" to track your .img files. Trust me, in six months you won't remember if Backup_Final_v2.img is the good one or the broken one.

## PART 1: THE CHIRP WORKFLOW CARD (CUT OUT)

### THE GOLDEN RULES OF PROGRAMMING

**1. CABLE CHECK:** Plug USB in first. Check Device Manager. No Yellow Triangle? Good.

**2. PHYSICAL CONNECTION:** Turn Radio OFF. Push K1 plug hard until it SNAPS. Turn Radio ON (100% Volume).

**3. READ FIRST:** Radio $\rightarrow$ Download from Radio. *Always do this first.*

**4. SAFETY SAVE:** File $\rightarrow$ Save As $\rightarrow$ Factory_Backup _[Date].img. *Never touch this file again.*

**5. EDIT:** Import from RepeaterBook, copy/paste from other tabs, or manual entry.

**6. WRITE:** Radio $\rightarrow$ Upload to Radio.

**7. VERIFY:** Unplug radio. Turn OFF/ON. Press VFO/MR to Channel Mode. Check Channel 1.

## PART 2: PROGRAMMING BACKUP LOG

*Track your file versions here. Don't guess.*

**RADIO MODEL:** _________________  **SERIAL/ID:** _________________

**RADIO MODEL:** _________________  **SERIAL/ID:** _________________

**RADIO MODEL:** _________________  **SERIAL/ID:** _________________

**RADIO MODEL:** _________________  **SERIAL/ID:** _________________


## PART 3: THE "RESCUE USB" CHECKLIST

*Keep a USB thumb drive in your Go-Bag with these files. If the internet dies, you can still program.*

- [ ] **CHIRP Installer:** The latest .exe or .dmg file.

- [ ] **Prolific Driver:** The 3.2.0.0 Windows installer (for the cable).

- [ ] **Factory Image:** The original read from your radio.

- [ ] **Current Config:** Your latest "Known Good" .img file.

- [ ] **PDF Reference:** A copy of this book or the RepeaterBook PDF for your state.

# APPENDIX D

## RANGE TEST LOG & RESULTS TRACKER

### INSTRUCTIONS:

1. Print this sheet.

2. Grab a clipboard and a pen.

3. Conduct the test with a partner (Base Station + Rover).

4. **Goal:** Find the "Hard Deck" (the maximum reliable distance) in every direction.

### PART 1: TEST PARAMETERS

*Variable changes affect results. Record what you are using today.*

DATE: _________________ TIME: ________________

WEATHER: _______________________________ (Rain/Fog kills UHF!)

## PART 2: THE CARDINAL LOG

*Drive North, South, East, and West. Check in at landmarks. Use the **RST System** (Readability 1-5, Signal 1-9).*

- **Readability 5:** Perfect.

- **Readability 3:** Understandable but difficult.

- **Readability 1:** Unreadable.

**NORTH ROUTE**

**SOUTH ROUTE**

**EAST ROUTE**

**WEST ROUTE**

## PART 3: RESULTS TRACKER & ADJUSTMENTS

**MY "HARD DECK" (Reliable Radius): Test 1**

- **North:** _________ Miles

- **South:** _________ Miles

- **East:** _________ Miles

- **West:** _________ Miles

**MY "HARD DECK" (Reliable Radius): Test 2**

- **North:** _________ Miles

- **South:** _________ Miles

- **East:** _________ Miles

- **West:** _________ Miles

# APPENDIX E

## COMMS PLAN, DRILL CARDS & AAR

**INSTRUCTIONS:**

1. **The Wallet Card:** Fill out, fold, laminate, and put in every family member's wallet or backpack.

2. **The Drill Cards:** Cut these out. Draw one from a hat on a Saturday morning to gamify your practice.

3. **The AAR:** Keep a stack of these in your radio binder. Fill one out after every storm or major drill.

## PART 1: THE FAMILY COMMS WALLET CARD

*(Cut along the outer line. Fold in the middle to make a double-sided card)*

| SIGNAL PLAN & CONTACTS | CODES & PROTOCOLS |
|---|---|
| PRIMARY CH: | CODE: "GET OUT" (LEAVE NOW) |
| BACKUP CH: | PHRASE: |
| EMERGENCY: Channel 0 (National Call) | CODE: "DURESS" (SEND HELP) |
| | PHRASE: |
| MOM: | CODE: "SILENCE" (DANGER) |
| DAD: | PHRASE: |
| OUT OF AREA: | IF LOST: STOP MOVING. |
| | 1. Turn Radio ON. Vol UP. |
| RALLY POINT: | 2. Hold Button. Count "1". |
| | 3. Say: "I am [Name]. I am at [Place]." |

# PART 2: THE "TACTICAL DAD" DRILL CARDS

*(Cut these out. Laminate them. Use dry-erase markers to fill in specifics)*

## DRILL 1: RADIO HIDE-AND-SEEK (Signal Mapping)

- **Objective:** Teach kids about "Line of Sight" and signal quality.

- **Roles:** Base (Dad), Rover (Kid).

- **The Mission:** Rover hides within 3 blocks. Must check in every 2 minutes.

- **The Test:**

  - [ ] Rover transmits from a low spot (creek/ditch). Base reports signal quality.

  - [ ] Rover transmits from a high spot (slide/hill). Base reports signal quality.

  - [ ] Rover describes landmark. Base confirms.

- **Success Criteria:** Kid understands that *Height = Clarity*.

## DRILL 2: THE SCAVENGER HUNT (Pro-Words)

- **Objective:** Teach "Copy," "Wilco," and listening skills.

- **Roles:** Command (Mom/Dad), Team Alpha (Kids).

- **The Mission:** Find 3 objects. Command gives instructions via radio ONLY.

- **The Test:**

  - [ ] Command: "Target 1 is a pinecone. Over."

  - [ ] Team: "Wilco. Searching." ... "Target secured."

  - [ ] Command: "Copy. Target 2 is a red leaf."

- **Success Criteria:** Kids wait for the transmission to end before moving. No interrupting.

## DRILL 3: THE **ZOMBIE RUN (Silent Ops)**

- **Objective:** Teach volume discipline and whispering.

- **Roles:** Zombies (Dad), Survivors (Mom/Kids).

- **The Mission:** Survivors must get from Point A to Point B without the Zombie hearing them on the radio or in person.

- **The Rules:**

  - Use Low Power (1W).

  - Use Headphones/Earpieces if available.

  - Whisper transmissions.

  - Use Code Words for the Zombie location.

- **Success Criteria:** Survivors reach the "Safe Zone" undetected.

# PART 3: AFTER ACTION REVIEW (AAR) LOG

*Fill* this out after *every drill or real-world event. Be honest.*

**EVENT:** _______________________________________

**DATE:** ______________ **DURATION:** ______________

**WEATHER:** ______________________________

## HARDWARE FAILURES:

- [ ] Battery died unexpectedly on Unit: _______

- [ ] Antenna loose/broken on Unit: _______

- [ ] Belt clip failed on Unit: _______

ACTION PLAN: What to do differently next time?

# APPENDIX F

## MAINTENANCE SCHEDULE (MONTHLY/QUARTERLY)

### PART 1: THE MONTHLY "READY CHECK" (10 MINUTES)

*GOAL: ENSURE THE GEAR will turn on TODAY if needed.*

- [ ] **Voltage Check:** Turn on every radio. Hold 0.

  - *Standard:* If < 7.6V, charge it. If > 8.0V, it's good.

- [ ] **Power Bank Check:** Press the button on your USB Anker/power bricks. Do they show 3-4 lights? Top off if needed.

- [ ] **Physical Tightness:**

  - Twist the antenna. (They loosen in bags).

  - Wiggle the Belt Clip. (Screws back out over time).

  - Check the Volume Knob. (Is it bent/stiff?).

- [ ] **The "Kerchunk" Test:** Transmit on your local repeater for 1 second. Did you hear the tail? (Verifies the transmitter works).

### PART 2: THE QUARTERLY "DEEP CLEAN" (30 MINUTES)

*Goal: Prevent corrosion and update software. Do this at the start of every season (Spring, Summer, Fall, Winter).*

- [ ] **The AA Inspection (CRITICAL):** Open your backup "Clamshell" cases.

  - ○ *Check:* Are the batteries leaking white powder?

  - ○ *Action:* If leaking, throw them out and clean contacts with vinegar. If good, rotate them (put them in the TV remote) and buy fresh ones for the kit.

- [ ] **Contact Scrub:** Remove the main battery. Look at the gold contacts on the radio and the battery.

  - ○ *Action:* Take a **pencil eraser** and scrub the gold contacts until they are shiny. This removes invisible oxidation that causes power failures.

- [ ] **Cable Audit:** Inspect your hand mic and earpiece cables.

  - ○ *Check:* Look for fraying, especially where the wire enters the plug.

  - ○ *Action:* If insulation is cracked, wrap with electrical tape or replace.

- [ ] **Go-Bag Rotation:** Take the radio out of the bag. Turn it on. Make a test call.

  - ○ *Why:* Electronics hate sitting dormant. Capacitors need to charge.

- [ ] **Digital Backup:** Connect the radio to the laptop. Open CHIRP.

  - ○ *Action:* Download the current config. Save it as Backup_[Season]_[Year].img. This ensures you have a copy of any frequency changes you made on the fly.

## PART 3: THE ANNUAL "BATTERY AUDIT"

*Do this once a year (e.g., New Year's Day).*

- [ ] **Capacity Test:** Leave your radio on "Scanning" mode. Time how long it takes to die.

  - ○ *Standard:* If it dies in under 4 hours, the battery is end-of-life. Replace it.

- [ ] **Replace Disposables:** Replace all Lithium AA batteries in your backup kits, even if they aren't dead. Fresh batteries = Peace of mind.

# APPENDIX G

## THE "TACTICAL DAD" GLOSSARY (TERMS USED IN THIS BOOK)

- **70cm (70 Centimeters):** The UHF band (420-450 MHz). Great for penetrating buildings and urban environments. GMRS lives here.

- **2m (2 Meters):** The VHF band (144-148 MHz). Great for outdoors, hills, and longer range in open terrain.

- **CHIRP:** The free, open-source software used to program Baofeng radios via a computer. Much faster than using the keypad.

- **Coax (Coaxial Cable):** The thick black cable that connects a radio to an external antenna. Think of it like a garden hose for radio waves.

- **CTCSS (Continuous Tone-Coded Squelch System):** Often called a "PL Tone" or "Privacy Tone." A sub-audible hum sent with your voice. It acts like a key; if the receiving radio (or repeater) doesn't hear the tone, it won't open the speaker. **It is NOT encryption.**

- **Desense (Desensitization):** When a cheap radio goes "deaf" because it is near a strong signal source (like a pager tower). The radio turns down its sensitivity to protect itself and blocks out the weak signal you want to hear.

- **Duplex:** Using two different frequencies to communicate (one to listen, one to talk). This is how repeaters work.

- **Duty Cycle:** The ratio of time spent Transmitting vs. Receiving vs. Standing by. Transmitting burns battery 20x faster than listening.

- **Frequency Mode (VFO):** The mode where you type in a frequency manually (e.g., 1-4-6-5-2-0). Used for finding new signals.

- **GMRS (General Mobile Radio Service):** A "family-friendly" UHF radio service that requires a license ($35, no test) but allows higher power and repeaters.

- **Ground Plane:** A metal surface (like a car roof) underneath an antenna that reflects radio waves upward, drastically increasing range.

- **Kerchunk:** Pressing the PTT button for 1 second and releasing it without speaking, just to see if the repeater activates.

- **Line of Sight:** The rule that VHF/UHF signals travel in straight lines. If there is a mountain between you and the target, the signal stops.

- **Memory Mode (MR):** The mode where you scroll through saved channels (Channel 1, Channel 2). Used for daily operation.

- **Offset:** The mathematical difference between a repeater's "Listen" frequency and its "Talk" frequency (usually 0.6 MHz or 5.0 MHz).

- **PTT (Push-to-Talk):** The big button on the side of the radio. Hold to speak, release to listen.

- **Repeater:** A high-power automated radio on a tower/mountain that receives a weak signal and simultaneously re-transmits it at high power to cover a wide area.

- **Reverse:** A function (often the * key) that allows you to listen to the *input* frequency of a repeater to see if the person talking is close enough to talk directly.

- **RF (Radio Frequency):** The electromagnetic waves generated by the radio.

- **Simplex:** Talking directly from radio to radio on the same frequency. No repeater involved. Range is limited by terrain (1-3 miles).

- **SMA (Sub-Miniature Version A):** The threaded connector type on the top of the radio. Baofengs use **SMA-Male** on the radio (a pin) and require **SMA-Female** antennas (a hole).

- **Squelch (SQL):** A "Gate" that keeps the speaker quiet when no signal is present. It filters out static. Setting it too high makes the radio deaf; setting it too low makes it noisy.

- **SWR (Standing Wave Ratio):** A measurement of how efficiently an antenna radiates power. High SWR means the power is bouncing back into the radio (bad). Low SWR means the power is leaving the antenna (good).

- **Transceiver:** A device that can both **Trans**mit and **Receiv**e (like your UV-5R).

- **VHF/UHF:** **V**ery **H**igh **F**requency / **U**ltra **H**igh **F**requency. The two main "bands" the UV-5R operates on.

- **VOX (Voice Operated Exchange):** A setting that transmits automatically when you speak (hands-free). **Keep this OFF** to avoid hot-mic incidents.

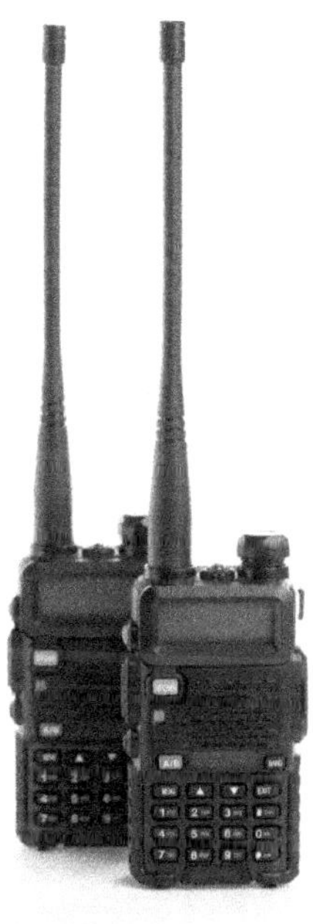